Roland Marc Bruderer

Regulation of Aurora B by Cdc48/p97 at the end of mitosis

Roland Marc Bruderer

Regulation of Aurora B by Cdc48/p97 at the end of mitosis

Cdc48/p97 promotes reformation of the nucleus by extracting the kinase Aurora B from chromatin

Südwestdeutscher Verlag für Hochschulschriften

Impressum/Imprint (nur für Deutschland/ only for Germany)
Bibliografische Information der Deutschen Nationalbibliothek: Die Deutsche Nationalbibliothek verzeichnet diese Publikation in der Deutschen Nationalbibliografie; detaillierte bibliografische Daten sind im Internet über http://dnb.d-nb.de abrufbar.
Alle in diesem Buch genannten Marken und Produktnamen unterliegen warenzeichen-, marken- oder patentrechtlichem Schutz bzw. sind Warenzeichen oder eingetragene Warenzeichen der jeweiligen Inhaber. Die Wiedergabe von Marken, Produktnamen, Gebrauchsnamen, Handelsnamen, Warenbezeichnungen u.s.w. in diesem Werk berechtigt auch ohne besondere Kennzeichnung nicht zu der Annahme, dass solche Namen im Sinne der Warenzeichen- und Markenschutzgesetzgebung als frei zu betrachten wären und daher von jedermann benutzt werden dürften.

Verlag: Südwestdeutscher Verlag für Hochschulschriften Aktiengesellschaft & Co. KG
Dudweiler Landstr. 99, 66123 Saarbrücken, Deutschland
Telefon +49 681 37 20 271-1, Telefax +49 681 37 20 271-0, Email: info@svh-verlag.de
Zugl.: Zürich, ETH, Diss. 2008

Herstellung in Deutschland:
Schaltungsdienst Lange o.H.G., Berlin
Books on Demand GmbH, Norderstedt
Reha GmbH, Saarbrücken
Amazon Distribution GmbH, Leipzig
ISBN: 978-3-8381-0785-1

Imprint (only for USA, GB)
Bibliographic information published by the Deutsche Nationalbibliothek: The Deutsche Nationalbibliothek lists this publication in the Deutsche Nationalbibliografie; detailed bibliographic data are available in the Internet at http://dnb.d-nb.de.
Any brand names and product names mentioned in this book are subject to trademark, brand or patent protection and are trademarks or registered trademarks of their respective holders. The use of brand names, product names, common names, trade names, product descriptions etc. even without a particular marking in this works is in no way to be construed to mean that such names may be regarded as unrestricted in respect of trademark and brand protection legislation and could thus be used by anyone.

Publisher:
Südwestdeutscher Verlag für Hochschulschriften Aktiengesellschaft & Co. KG
Dudweiler Landstr. 99, 66123 Saarbrücken, Germany
Phone +49 681 37 20 271-1, Fax +49 681 37 20 271-0, Email: info@svh-verlag.de

Copyright © 2009 by the author and Südwestdeutscher Verlag für Hochschulschriften Aktiengesellschaft & Co. KG and licensors
All rights reserved. Saarbrücken 2009

Printed in the U.S.A.
Printed in the U.K. by (see last page)
ISBN: 978-3-8381-0785-1

Table of contents

Summary 3
Zusammenfassung 4
1. Introduction 5
   1.1 Mitosis in animal cells 5
      1.1.1 Function of mitosis 5
      1.1.2 Phosphorylation in mitosis and Aurora B 8
      1.1.3 The ubiquitin system 10
   1.2 The AAA-ATPase p97 14
      1.2.1 p97 14
      1.2.2 Roles of p97 in interphase processes 17
      1.2.3 The roles of p97 in mitosis 20
   1.3 Functional interaction of p97 with Aurora B in *Xenopus laevis* egg extracts during nucleus formation 25
      1.3.1 Nuclear envelope formation in *Xenopus laevis* egg extract 26
      1.3.2 Discovery of the $p97^{Ufd1-Npl4}$ Aurora B functional interaction 27
   1.4 Aim of this thesis 29
2. Results 30
   2.1 *In vitro* reconstitution of nuclear envelope formation and ubiquitination 30
      2.1.1 Preparation of crude and fractionated *Xenopus laevis* interphase extract 30
      2.1.2 *In vitro* nucleus formation and quantification 31
      2.1.3 *In vitro* ubiquitination analysis using *Xenopus laevis interphase* egg cytosol 33
   2.2. The first evidence for a role of the chromosomal passenger complex in nucleus formation 33
   2.3 Interaction analysis of p97 with the chromosomal passenger complex 38
   2.4 Inactivation of p97 results in accumulation of ubiquitinated Aurora B and survivin 42
   2.5 Chromosomal passenger complex fate in *Xenopus laevis* egg interphase cytosol 45
   2.6 Chromosomal passenger complex association with chromatin 46
   2.7 Mobilization of Aurora B from chromatin 47
      2.7.1 Establishment of the Aurora B Mobilization assay 48
      2.7.2 Mobilization of Aurora B from chromatin is dependent on p97 49
      2.7.3 Ubiquitinated Aurora B accumulates on chromatin in absence of p97 function 53
   2.8 Mobilization of Aurora B from chromatin influenced by proteasome activity 54
   2.9 Interaction of the ligase Cullin3 with Aurora B and p97 55
   2.10 Analysis of polyubiquitination of Aurora B by Cullin3 56
      2.10.1 Analysis of ubiquitination of Aurora B by Cullin3 using immunodepletion 56
      2.10.2 Analysis of polyubiquitination of Aurora B by Cullin3 using depletion with a fragment of KLHL13 57
      2.10.3 Analysis of polyubiquitination of Aurora B by Cullin3 using a dominant negative KLHL13 fragment 61
   2.11 Regulation of the ATPase activity of p97 by p47 and Ufd1-Npl4 62
3. Discussion 67
   3.1 Functional Role of p97 in nucleus formation 67
   3.2 $p97^{Ufd1-Npl4}$ does not mediate membrane fusion during NEF 68
   3.3 p97 is a negative regulator of Aurora B 69
   3.4 p97 physically extracts Aurora B from chromatin 69

3.5 Aurora B is a novel inhibitor/regulator of nucleus formation 74
3.6 Fate of Aurora B after extraction from chromatin 76
3.7 A Cullin3-based ligase complex may cooperate with p97 77
3.8 General relevance on p97, Cullin3 and Aurora B in mitosis 79
4. Material and Methods 81
4.1 Preparation of crude and fractionated *Xenopus laevis* egg extract 81
4.2 Preparation of demembranated sperm chromatin 82
4.3 Nuclear envelope formation assay 83
4.4 Ubiquitination analysis 83
4.5 Depletions and immunoprecipitations 83
4.6 Aurora B mobilization assay 84
4.7 Analysis of chromatin bound ubiquitinated proteins 84
4.8 Malachite green ATPase Assay 84
4.9 Protein expression 85
4.10 SDS-PAGE and Western blotting 89
4.11 Antibodies 90
4.12 Affinity purification of anti-survivin antibodies 90
4.13 Purification of anti-INCENP antibodies 91
4.14 Cloning 91
5. References 93
Abbreviations 105
Acknowledgement 108

Summary

During mitosis, the replicated genome is segregated into two daughter nuclei. It is the most dramatic phase of the cell cycle, as chromatin condenses into chromosomes followed by the breakdown of the whole nucleus. After sister chromatid segregation by the mitotic spindle, the nuclei reform in the daughter cells and cell division can be completed with cytokinesis. Chromatin needs to be segregated and repackaged into nuclei with high fidelity, since errors cause genomic instability that in turn may lead to cell death or, worse, development of cancer. Reformation of the nucleus late in mitosis needs a high degree of coordination with chromosome segregation and spindle disassembly, but also among the NE formation, nuclear pore complex assembly and chromatin decondensation; however, regulation of the process is still poorly understood. *In vitro*, nucleus formation requires p97, a hexameric ATPase implicated in membrane fusion and ubiquitin-dependent processes. However, the role and relevance of p97 in nucleus formation have remained controversial.

Here we show that p97 stimulates nucleus reformation by inactivating the chromatin-associated kinase Aurora B. During mitosis, Aurora B inhibits nucleus reformation by preventing chromosome decondensation and formation of the nuclear envelope membrane. During exit from mitosis, p97 binds to Aurora B after its ubiquitination and extracts it from chromatin. This leads to inactivation of Aurora B on chromatin, thus allowing chromatin decondensation and nuclear envelope formation.

These data reveal an essential pathway that regulates reformation of the nucleus after mitosis and defines ubiquitin-dependent protein extraction as a common mechanism of $p97^{Ufd1-Npl4}$ activity also during nucleus formation.

Zusammenfassung

Während der Mitose wird das replizierte Genom in die zwei Tochterzellen segregiert. Dies ist eine dramatische Phase des Zellzyklus, das Chromatin kondensiert zu Chromosomen und der Zellkern zerfällt. Nachdem die mitotische Spindel die Schwesterchromatide segregiert hat, bilden sich neue Zellkerne in den Tochterzellen und die Zellteilung kann mit der Zytokinese zum Abschluss gebracht werden. Chromatin muss mit höchster Präzision segregiert und wieder in Zellkerne verpackt werden, da Fehler genetische Instabilitäten verursachen und zum Zelltod, oder schlimmer, zur Bildung von Krebs führen können. Die Neubildung des Zellkerns, spät in der Mitose, erfordert ein hohes Mass an Koordination mit der Chromosomen Segregation und dem Zerfall der Spindel. Auch die Bildung einer neuen Kernmembran, der Aufbau der Kernporen und die Dekondensation des Chromatins benötigen Koordination. Es ist jedoch wenig über die Regulation dieser Prozesse bekannt. *In vitro* ist die Bildung des Zellkerns von p97, einer hexameren ATPase, abhängig, die in Membranfusion und in Ubiquitin abhängige Prozesse involviert ist. Die Rolle und Relevanz von p97 während der Zellkernbildung waren bisher allerdings kontrovers diskutiert worden.

Mit dieser Arbeit zeigen wir, dass p97 die Bildung des Zellkerns stimuliert, indem es die Chromatin gebundene Kinase Aurora B inaktiviert. Während der Mitose inhibiert Aurora B die Kernbildung, indem sie die Dekondensation des Chromatins und die Bildung einer Kernmembran verhindert. Am Ende der Mitose bindet p97 an Aurora B nach deren Ubiquitinierung und entfernt Aurora B von Chromatin. Dies führt zur Inaktivierung von Aurora B auf dem Chromatin, und erlaubt daher die Dekondensation des Chromatins und die Bildung einer Kernmembran.

Diese Daten decken einen essenziellen Regulationsweg auf, der die Bildung des Zellkerns nach der Mitose steuert, und sie definieren Ubiquitin-abhängige Proteinextraktion als grundlegende Aktivität von $p97^{Ufd1-Npl4}$ auch während der Bildung des Zellkerns.

1. Introduction

1.1 Mitosis in animal cells

1.1.1 Function of mitosis

Cell division is a very fundamental process of cellular life. It is vital for the persistence of life and evolution. It provides the inheritance of the genetic material from generation to generation and is the foundation of evolution. In eukaryotic cells, two types of cell division can occur. In a cell division with mitosis, a cell duplicates its genetic material and separates the chromosomes, into two identical sets in two daughter nuclei. This process results in two cells with a diploid number of chromosomes. In contrast, a cell division with meiosis is essential for sexual reproduction and therefore occurs in all eukaryotes that reproduce sexually. In a meiotic cell division, the genome of a diploid germ cell undergoes replication of the genetic material followed by two rounds of division, resulting in four haploid cells.

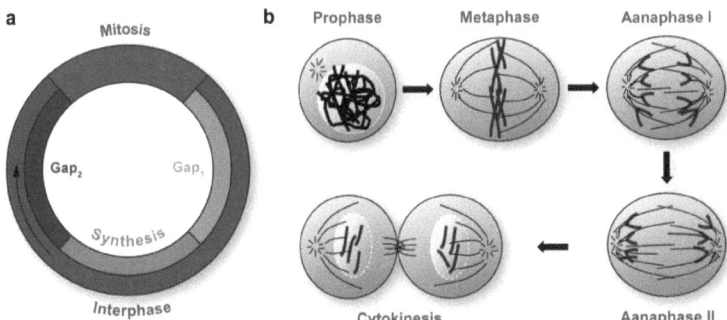

Fig. 1 Cell cycle and mitosis. a, mitosis is a part of the cell cycle. In the interphase a cell prepares itself for mitosis. After Gap_1 with one set of unreplicated chromatin, the S-phase starts. The chromatin is replicated resulting in duplicated set of chromosomes. Subsequently, Gap_2 starts after the genetic material is duplicated. After Gap_2, interphase ends and mitosis starts. **b**, in prophase the chromatin condenses to high order structures called chromosomes, this results paired sister chromatids. The mitotic spindle starts to form. At nuclear envelope breakdown prometaphase starts. The chromosomes become attached to the spindle and align in the 'metaphase plate'. During anaphase, the chromosome pairs segregate (I) and the spindle elongates (II). In telophase, the cell starts processes that result in cytokinesis. At the same time the chromatin decondenses and the daughter nuclei form. After successful cytokinesis, a round of mitosis is completed.

The process of mitosis was discovered in the 1800s and by 1950 the basic events of the process had been documented by light microscopists. Substantial rearrangements and changes of the morphology of the cell take place. These changes are easily visible by light microscopy and were distinguished into different phases termed the prophase, metaphase, anaphase and telophase.

The phases of mitosis

The mitotic phase is only a part of the complete cell cycle. For a successful mitosis, a cell must have passed successfully through a phase of preparation, called interphase. Interphase can be divided into three phases G_1, S, G_2, plus a "post-mitotic" phase G_0. During G_1 (Gap 1) cells grow and prepare for the next phase. During S phase, cells replicate DNA and duplicate their whole genome. Then in G_2 (Gap 2), cells prepares themselves for mitosis, which involves significant protein synthesis (Figure 1a). If a cell decides to go into mitosis, the interphase chromatin condenses into highly ordered structures called chromosomes. The duplicated sister chromosomes are linked one to the other. This phase is called the prophase. The mitotic spindle starts to form late in prophase. Prophase ends, when the nuclear envelope breaks down. Cellular organelles like the Endoplasmic Reticulum and Golgi stacks are reorganized and/or fragmented, allowing distribution in the two resulting daughter cells at the end of mitosis (Shorter and Warren 2002; Uchiyama and Kondo 2005). In prometaphase, stable microtubules asters emerge from the spindle pole bodies and attach to kinetochores on the chromosomes using a search and capture mechanism. After successful connection of the spindle to the chromosomes, metaphase begins and the chromosome pairs are aligned along the 'metaphase plate', which is perpendicular to the spindle axis. Once this is achieved, anaphase starts. The sister chromatids are detached and segregated rapidly and synchronously (anaphase I). Subsequently, the spindle elongates (anaphase II). At this stage a structure forms in the center of the spindle consisting of antiparallel microtubules bundles and structural and regulatory proteins. This structure is called midzone or central spindle. When the chromatids have reached the opposite spindle poles, the chromatin starts to decondense and the spindle disassembles. In telophase, a nuclear envelope reforms and encloses the decondensing chromatin masses forming daughter nuclei. During late anaphase and telophase, the cell begins to divide. An actomyosin contractile ring constricts the plasma membrane to generate two daughter cells connected by a cytoplasmic bridge. The spindle midzone is compressed and now termed midbody. Finally the cytoplasm is partitioned physically into two compartments and the two daughter cells become independent (Glotzer 2005). The completion of cytokinesis marks the end of mitosis (reviewed in (McCollum 2005), (Pines 2006)) (Figure 1b).

Regulation of mitosis

Regulation of all these mitotic processes requires precise regulation and is highly important for the cell. Small mistakes during mitosis often have deleterious effects resulting in genetic instability, cell death or transformation. Molecular details of mitosis that ensure the enormous fidelity of this process have been discovered and described only later. Entry into mitosis is achieved by the activation of the mitotic cyclin-dependant kinases, which regulate a variety of events by phosphorylation. Cyclins oscillate during mitosis in a regulated manner. Cyclin destruction is mediated by the anaphase-promoting complex / cyclosome APC/C. The APC/C is an ubiquitin ligase that targets specific key proteins for degradation during mitosis in temporal controlled manner. The attachment of the chromosomes to the microtubule asters of the mitotic spindle is controlled by the spindle assembly checkpoint, a control mechanism that ensures correct bipolar attachment and corrects monotelic, syntelic and merotelic attachments. The spindle assembly checkpoint is inactivated once all chromosome pairs are correctly attached to microtubules (Tan, Rida et al. 2005), and the APC/C initiated dephosphorylation by inducing degradation of Cyclin B. The irreversible separation of the two sister chromatids at the metaphase to anaphase transition is regulated by the protease separase. Separase is activated after the spindle assembly checkpoint is inactivated, it cleaves the Scc1 unit of cohesin, which links the sister chromatids together. Separase is regulated by several mechanisms, the best understood is the degradation of the separase inhibitor securin by the APC/C (Uhlmann 2003). Exit of mitosis in yeast is regulated by the mitotic exit network (MEN) and the CDC fourteen early anaphase release (FEAR) network (Dumitrescu and Saunders 2002).

Two predominant underlying mechanisms are ensuring the regulation of faithful and directional mitosis namely protein phosphorylation and regulated protein degradation. These regulation mechanisms are of great importance for cells since errors in the choreography of the processes in mitosis lead to genetic instability or aneuploidy, possibly resulting in cell death or disease.

1.1.2 Phosphorylation in mitosis and Aurora B

Mitotic kinases

The cyclin dependent kinases are the mayor regulators of the cell cycle. They guide the cell through the cell cycle and mitosis by regulating a huge variety of proteins by phosphorylation. The different processes are switched on and off at precise times and locations throughout the cell. The regulated processes range from the DNA replication, centrosome duplication, phosphorylation to disassembly of the nuclear pore complexes (Margalit, Vlcek et al. 2005) or the activity of the anaphase promoting complex. Cyclin dependent kinase activity is regulated by the cyclins. Cyclins are proteins that have a defined lifetime and activity during stages of the cell cycle (reviewed in (Murray 2004)). The cyclin dependent kinases work in concert with other mitotic kinases like the Aurora kinases, POLO, NIMA families and others in a network. The Aurora class consists of Aurora A that is involved in centrosome function, mitotic entry and spindle assembly and Aurora B that participates in chromosome modification, correct attachment of chromosomes to microtubules as well as cytokinesis, and Aurora C that is partially redundant to Aurora B and its function remains elusive (Fu, Bian et al. 2007). POLO kinases are important for centrosome maturation and activation of the APC/C^{cdc20}. NIMA kinases cooperate with the cyclin dependent kinase1 cdk1 at the G_2/M transition (reviewed in (Nigg 2001)). Many of these kinases are regulated by the cyclin dependent kinases.

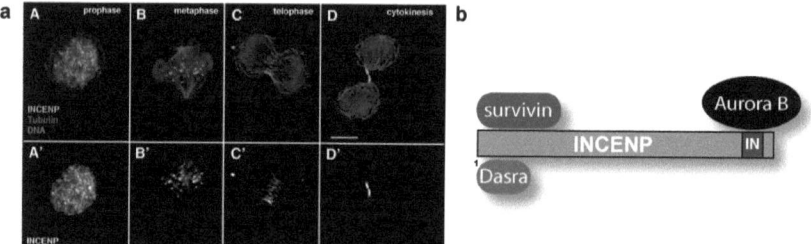

Fig. 2 Chromosomal passenger complex localization during mitosis. a, the chromosomal passenger complex comprises the proteins Aurora B, INCENP, survivin, TD60, and Dasra/Borealin B/Csc-1. The localization of INCENP is shown. The chromosomal passenger complexes associated with the chromatin masses during prophase. In metaphase, it localizes to the inner centromeres and then appears at the central spindle and lateral equatorial cortex in anaphase. Finally, the chromosomal passenger complex is localized at the midbody during cytokinesis. (images taken from (Vagnarelli and Earnshaw 2004)). **b**, Scheme of the interactions within the chromosomal passenger complex.

The mitotic kinases are antagonized by phosphatases; a controlled balance between them exists. Phosphorylation and dephosphorylation is a rapid and reversible regulation. But it can become irreversible, when a kinase or phosphatase is degraded. The phosphatases are also regulated during mitosis (Trinkle-Mulcahy and Lamond 2006). Important mitotic phosphatases are protein serine/threonine phosphatases PP1 and PP2A and the tyrosine phosphatases CDC25 and Cdc14 in yeast.

The chromosomal passenger complex kinase Aurora B

In this thesis, the mitotic kinase Aurora B is a central importance, therefore, a more detailed introduction is provided here. Aurora B is a member of the chromosomal passenger complex. The chromosomal passenger complex has its name from its remarkable localization during mitosis. Early in mitosis, it is associated with the chromatin masses. At prometaphase and metaphase it is located at inner centromeres. Next, it appears at the central spindle and lateral equatorial cortex in anaphase. Finally, the chromosomal passenger complex is localized at the midbody during cytokinesis (Figure 2a). Other passenger proteins are the inner centromeric protein INCENP, survivin, TD60, and Dasra/Borealin B/Csc-1 (Vader, Medema et al. 2006). INCENP may function like a binding platform for the other proteins. Survivin and Dasra bind to the N-terminus of INCENP, Borealin and INCENP associate with the helical domain of survivin to form a tight three-helical bundle (Jeyaprakash, Klein et al. 2007). Dasra promotes binding of survivin to INCENP (Vader, Kauw et al. 2006). Aurora B binds tightly to the C-terminal IN-Box of INCENP (Figure 2b). This interaction is required to activate Aurora B kinase activity, as well as phosphorylation on Threonine 248 of Aurora B in *Xenopus laevis* (Sessa, Mapelli et al. 2005), although it is not always sufficient for activation of Aurora B *in vivo* (Yamamoto, Lewellyn et al. 2008). In addition to that, clustering of the Aurora B kinase is required for full activation (Kelly, Sampath et al. 2007). Additionally, also survivin can activate the Aurora B kinase activity (Honda, Korner et al. 2003), (Chen, Jin et al. 2003). Protein phosphatase 1 can be a negative regulator of Aurora B by preventing accumulation of phosphorylation on Aurora B (Sugiyama, Sugiura et al. 2002) and (Murnion, Adams et al. 2001). Survivin is a conserved member of the inhibitor of apoptosis protein family. The baculovirus IAP repeat domain of survivin is responsible for its dimerisation. The function of Dasra is less understood. It may be required together with survivin for chromatin and centromere targeting of the chromosomal passenger complex (Kelly, Sampath et al. 2007), (Vader, Kauw et al. 2006). The function of TD60 remains elusive; very recently it was suggested that TD60 is required to localize the chromosomal passenger complex to centromeres and to coordinate with microtubules to activate the kinase activity of Aurora B (Rosasco-Nitcher, Lan et al. 2008). The chromosomal passenger complex has many important functions during mitosis; it is involved in the establishment of a stable bipolar spindle and corrections of kinetochore-microtubule attachment errors through sensing the lack of tension and activation of the spindle assembly checkpoint. Additionally, it is involved in sister chromatid cohesion, regulation of mitotic progression and cytokinesis (reviewed in (Ruchaud, Carmena et al. 2007)). Aurora B and survivin are degraded at the end of mitosis (Honda, Korner et al. 2003). In some systems, Aurora B is degraded in an APC/C^{cdh1} dependent fashion at the end of mitosis (Stewart and Fang 2005), (Nguyen, Chinnappan et al. 2005). Precise regulation of the kinase activity of Aurora B is crucial as too low or too high activity is deleterious for cells and can result in genetic instability (Li and Li 2006).

1.1.3 The ubiquitin system

Directionality in mitosis is achieved by controlled degradation of specific proteins at specific locations at specific times points. In eukaryotic cells, the accurate local and temporal degradation of proteins is executed by the ubiquitin system. The central components of the ubiquitin system are a small conserved 76 amino acid protein, called ubiquitin, and often a large multiprotein complex, the 26S proteasome that degrades the substrate proteins. Besides the progression through the cell cycle, the ubiquitin system is also involved in transcription activation, in postreplicational DNA repair, in ribosomal function, in the induction of the inflammatory response, in antigen presentation and more processes (Pickart 2001). Ubiquitin is attached to substrate proteins of the ubiquitin system. For degradation, ubiquitin chains are generated on the substrate protein. The multiubiquitin chains on the substrate are recognized by processing and escorting factors including the ubiquitin selective chaperone p97. Finally, the substrate is handed to the proteasome, where degradation of the substrate protein into short peptides and recycling of the ubiquitin moieties occurs (reviewed in (Hershko and Ciechanover 1998)). The proteasome is constitutively active throughout the cell cycle. Therefore, regulation is achieved at the stage of ubiquitination of the substrate.

The enzymatic cascade E1, E2 and E3
An enzymatic cascade is responsible for the selective ubiquitination of substrate proteins. This cascade consists of three layers of enzymes. The ubiquitin-activating enzymes (E1) are the first layer; only one gene of the E1 with two splice variants exists in mammals. Ubiquitin-conjugating or ubiquitin-carrier enzymes (E2) are the second layer; about two dozen exist in

mammals. The ligases are the layer, which mediate substrate specificity and which exist in a multiplicity. The cascade of ubiquitination starts with the activation of ubiquitin. Free ubiquitin is adenylated at the C-terminus glycine by the E1. This high-energy intermediate is subsequently transferred to the active site cysteine of the E1, creating a thioester linkage and releasing AMP. Subsequently, ubiquitin is transferred to the catalytic cysteine of the E2 via a transthiolation reaction. Thirdly, the ubiquitin is transferred to the ε-amino group of a lysine side chain on the surface of the substrate forming an isopeptide bond. This step is catalyzed by ubiquitin ligase. The ligases confer the substrate specificity in the UPS system. Two main classes of ubiquitin ligases exist based on structural and mechanistic features. The first class is the RING ubiquitin ligases containing the RING motif. They function as bridging factors between the ubiquitin conjugating enzyme and the substrate, allowing specific ubiquitination. RING ligases do not form a covalent bond to the ubiquitin moiety and are often multimeric protein complexes. Prominent examples are the anaphase-promoting complex / cyclosome (APC/C) and the cullin-based ligases (Ozkan, Yu et al. 2005). The second class of ubiquitin ligases contains a HECT domain. The ubiquitin moiety is transferred from the ubiquitin-conjugating enzyme to a conserved cysteine in the HECT domain forming a thioester linkage. Subsequently, ubiquitin is transferred to the substrate (Ogunjimi, Briant et al. 2005).

This process can be executed in multiple rounds and thereby ubiquitin chains can be generated on the substrate. The chain formation is achieved by attachment of the C-terminal residue glycine of one ubiquitin molecule linked through an isopeptide bond to a lysine residue within another (Figure 3).

Ubiquitin modifications

The following types of modifications occur *in vivo* mono-ubiquitination, multiubiquitination and polyubiquitination chains. Ubiquitin chains can be formed using all the lysines of ubiquitin at lysine 6, 11, 27, 29, 33, 48 and 63. Additionally, also branched polyubiquitin chains exist. The lysine-48 chains are the most abundant followed by the lysine-63 chains (Peng, Schwartz et al. 2003). The linkage of the chains encodes additional signals. Lysine-48 chain ubiquitination with at least 4 ubiquitin moieties is sufficient to target a substrate protein for proteasomal degradation. If a substrate protein is ubiquitinated at one lysine with one ubiquitin moiety, it is mono-ubiquitinated. This modification has a variety of consequences, not including proteasomal degradation. For example, mono-ubiquitination regulates the activity of transcription factors in the nucleus (Sloper-Mould, Jemc et al. 2001). In yeast, it has been shown that attachment of an ubiquitin moiety to a model transcription factor is required for normal transcriptional activity. If a protein is modified at multiple lysines with one moiety of ubiquitin each, it is multiubiquitinated. Receptor tyrosine kinases become multiubiquitinated following ligand stimulation. Multiubiquitination acts as a signal controlling receptor internalization and direction to the lysosome for destruction, thereby leading to receptor downregulation (Haglund, Di Fiore et al. 2003; Mosesson, Shtiegman et al. 2003). For other proteins, specifically a subset of yeast nutrient permeases, maximal internalization rates require modification with di-ubiquitin chains that are lysine-63-linked (Galan and Haguenauer-Tsapis 1997). In addition to that, lysine-63 linkage has been reported to be associated with neurodegenerative diseases. It promotes the formation and autophagic clearance of protein inclusions (Tan, Wong et al. 2008).

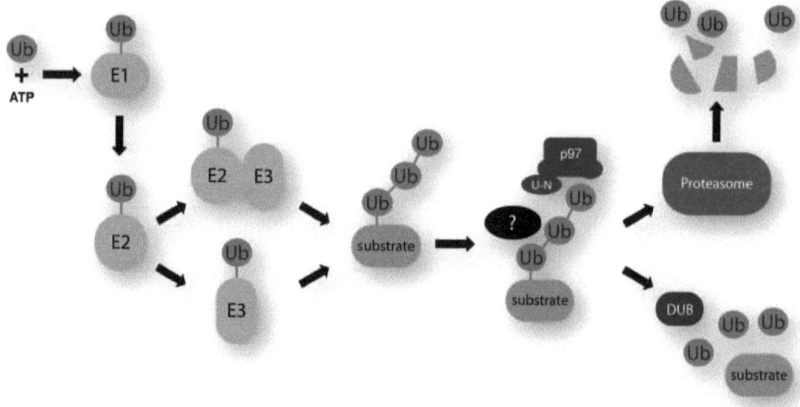

Fig. 3 Schematic overview of the ubiquitination pathway. Free ubiquitin is adenylated at the C-terminus by the E1. This high energy intermediate is subsequently transferred to the active site cysteine of the E1, creating a thioester linkage and releasing AMP. Subsequently, ubiquitin is transferred to the catalytic cysteine of the E2 via a transthiolation reaction. Finally, E3 ligases catalyze the ligation of the ubiquitin to its substrate protein to the ε-amino group of a lysine side chain on the surface of the substrate forming a isopeptide bond. In the case of HECT E3 ligases, the ubiquitin is linked to the active site cysteine in the HECT domain via a thioester bond prior to conjugation to the substrate. The ligase brings the E2 and the substrate in close proximity in the case of RING finger E3 ligases. Then, the ubiquitin is directly transferred from the E2 to the substrate. After ubiquitination, additional factors recognize and interact with the substrate. They edit, escort, shield or remodel the substrate. p97 is an example of a factor cooperating with the ubiquitin system. Finally, the substrate is either degraded or deubiquitinated.

Ubiquitin chain editing

An additional layer of regulation in the ubiquitin system is the editing of ubiquitin modifications. Deubiquitinating enzymes revert mono-ubiquitinated proteins to the non-modified state and polyubiquitin chains can be shortened, edited or eliminated. This may result in rescuing substrate proteins from proteasomal degradation, changing the activity of substrate proteins or recycling the ubiquitin moieties before proteasomal degradation of substrate proteins. Deubiquitinating enzymes at the proteasome are likely to have little substrate specificity, since a variety of substrates are deubiquitinated. In addition to that, several deubiquitinating enzymes have been discovered exhibiting substrate specificity. Deubiquitination is also important for the homeostasis of the ubiquitin levels. Inactivation of deubiquitinating enzymes results in retarded proteasomal degradation (Swaminathan, Amerik et al. 1999).

Ubiquitin binding domains
Recently there have been many factors discovered that contain ubiquitin binding domains. These factors interact with ubiquitinated proteins and can prevent the conversion of mono-ubiquitination into polyubiquitination, or protect from deubiquitinating enzymes. Furthermore, they may escort substrates to the proteasome or mediate downstream signalling. At least sixteen domain families were identified with various grades of affinity and specificities (Hurley, Lee et al. 2006). Various proteins that are neither part of the ubiquitination cascade nor part of the proteasome but that contain ubiquitin binding domain were discovered. Dsk2 and Rad23 are homologue ubiquitin recognizing molecules delivering substrate proteins to the proteasome. Therefore, they contain an ubiquitin associating (UBA) domain, binding lysine-48 chains and at an ubiquitin like domain, which binds to the proteasome (Funakoshi, Sasaki et al. 2002). A conserved vacuolar protein sorting pathway controls the trafficking of proteins to the vacuole/lysosome. The factor Vps9p, which promotes endosomal and Golgi-derived vesicle fusion, binds to ubiquitin via a CUE ubiquitin binding domain. The CUE domain is essential for the trafficking (Davies, Topp et al. 2003). PCNA (proliferating cell nuclear antigen) is a cofactor of DNA polymerases that encircles DNA and orchestrates the function of the polymerases by recruiting crucial players to the replication forces. PCNA mono-ubiquitination recruits the translesion synthesis polymerases to stalled replications at DNA lesions (Moldovan, Pfander et al. 2007).
A conserved, essential and very abundant protein is p97; it acts in the interface between the ubiquitination cascade and substrate degradation or procession. It was reported to bind ubiquitin itself and associate with many proteins containing ubiquitin binding sites (reviewed in (Ye 2006)). Additionally, p97 cooperates with ligases as well as deubiquitinating enzymes. Hence, it appears to be a central factor coordinating the processes during ubiquitination by connecting ligases to substrates, bringing in additional factors to edit or shorten ubiquitin

chains in a variety of ubiquitin dependent cellular processes. It is of central importance to understand how p97 integrates and regulates these processes.

1.2 The AAA-ATPase p97

1.2.1 p97

p97 was originally discovered as a gene that contains the peptide Valosin. Therefore, p97 is also called Valosin-containing protein VCP in the mammals. Valosin had been isolated from pig intestine (Schmidt, Mutt et al. 1985) and was shown to affect gastro-intestinal activity (Konturek, Schmidt et al. 1987). Years later, p97 was rediscovered as a ubiquitous ATPase in *Xenopus laevis* egg extracts and named p97 for its relative molecular mass of 97 kDa (Peters, Walsh et al. 1990).

p97 is a conserved, essential and very abundant protein. p97 is about 1% of the cytosolic protein mass thereby being highly overrepresented, since in the human genome about 85000 human transcription units and proteins are predicted (Davison and Burke 2001). p97 is evolutionary conserved from yeast, where it is called CDC48 in *Saccharomyces cerevisiae*, to human with an identity (67.3%) and homology (14.7%) of total 82%. It is involved in a variety of cellular processes including the Endoplasmic Reticulum associated degradation, transcription activation (see **1.2.2**), mitotic processes ranging from Golgi reformation to spindle disassembly to nucleus formation (see **1.2.3**), and apoptosis. p97 is considered to be an ubiquitin dependent chaperone.

p97 structure and catalytic activity

p97 is a member of the large family of AAA ATPases (ATPases associated with a variety of cellular activities). The AAA protein family is subfamily of the AAA+ ATPases and belongs to the Walker-type NTPases. The proteins of this family are classified by the presence of one or more AAA domains (Ogura and Wilkinson 2001). The AAA domain contains the Walker A motif (GX4GKT) for nucleotide binding, Walker B motive (HyDE) for nucleotide hydrolysis and the second region of homology (SRH). AAA proteins are involved in a variety of cellular processes ranging from recombination, proteolysis, transcriptional activation, organelle biogenesis to cell cycle regulation (Snider and Houry 2008). The characteristic of AAA proteins is the coupling of chemical energy of ATP hydrolysis, with conformational changes usually to exert mechanical force on substrates. Thereby, they unfold substrates or disassemble multi-protein complexes and aggregates.

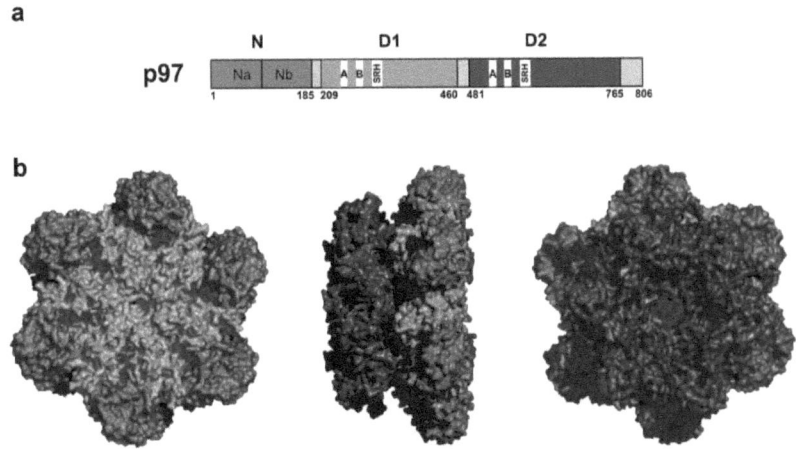

Fig. 4 Domain outline and crystal structure of a p97 hexamer. a, Outline of the domain structure of p97. The N domain binds the adaptor proteins. D1 and D2 are the conserved AAA ATPase domains with the Walker A and B motifs for ATP binding and hydrolysis, respectively. Additionally, D1 and D2 contain the second region of homology (SRH). **b,** Depicted is from left to right a top, side and bottom view of the full length crystal structure of p97. The D1 domain is shown in cyan, the D2 in darker blue and the N-domain in green. p97 forms a hexameric structure in which the D1 and D2 domains are stacked, while the N-domain lies outside of the barrel in plane with the D1 domain (Huyton, Pye et al. 2003).

p97 contains two AAA domains termed D1 and D2 and in addition to that an N-terminal domain called N-domain. p97 is a homohexamer in its native conformation (Figure 4a). The hexameric state is not dependent on ATP binding (Wang, Song et al. 2003). The AAA domains form a ring-shaped barrel and the two AAA-domains of a protomer are stacked. The resulting pore in the middle of the six fold symmetric ring is narrow and it was reported in a structure reported to bind a Zn^{2+} ion giving rise to speculation that the substrates of p97 are not threaded through the pore as it its reported for other AAA ATPases (DeLaBarre and Brunger 2003). The N-domain consists of two sub domains (Na and Nb) separated by a cleft. The N-domain is located in plane with D1 in the periphery of the hexamer (Huyton, Pye et al. 2003). The crystal structure of p97 is depicted in (Figure 4). The D2 domain of p97 is active in ATP hydrolysis. The ATPase activity of D2 is essential for substrate remodelling. Nucleotide binding in D1 is required for substrate binding (as shown in ER associated degradation

(ERAD)). The ATPase activity of D1 is low, one explanation for this is the high affinity for ADP of D1 (Briggs, Baldwin et al. 2008). The N-domain binds cofactor proteins of p97 and is reported to have regulatory functions on p97 ATPase activity (Rothballer, Tzvetkov et al. 2007) and to be mobile to a certain degree (Beuron, Dreveny et al. 2006).

Substrate recruiting and processing factors system of p97
As mentioned above, p97 is involved in a variety of different cellular processes with use of different accessory proteins. This led to the formulation of the adaptor hypothesis. p97 is thought to achieve its selectivity for its substrates in different processes through the use of substrate specific adaptors. Characterized adaptors are the heterodimeric Ufd1-Npl4, p47 and p37. These adaptor proteins contain a UBX or UBD domain that binds to the N-domain of p97. The UBX domain emerges to be a conserved p97-binding domain. Recently, many other proteins containing a UBX domain have been discovered, these all are putative substrate adaptors of p97 (Buchberger, Howard et al. 2001) (Schuberth and Buchberger 2008). p37 and p47 bind p97 with the UBX domain and additionally with a second binding motif called BS1.The heterodimeric Ufd1-Npl4 contains a UBX/UBD domain in Npl4 and a second binding motif BS1 on Ufd1 for p97 (Bruderer, Brasseur et al. 2004).

p37 and p47 direct p97 function to ER and Golgi biogenesis, p47 at the end of mitosis for Golgi reformation, p37 is involved in interphase (Uchiyama and Kondo 2005), (Uchiyama, Totsukawa et al. 2006), (Meyer, Wang et al. 2002). Additionally, p47 is required for growth of the nucleus, after nuclear envelope sealing (Hetzer, Meyer et al. 2001).Ufd1-Npl4 connects p97 in some processes to the ubiquitin proteasome system. Examples are p97$^{Ufd1-Npl4}$ function in ER associated degradation (Hampton 2002), nuclear envelope formation (Hetzer, Meyer et al. 2001) or transcription factor activation (Rape, Hoppe et al. 2001).

Beside the 'classical' adaptor proteins containing a UBX/UBD domain, a multitude of other p97 interacting proteins have been identified with about half a dozen different binding domains and motifs identified so far (Figure 5). These proteins comprise recruiting and processing factors of p97 complexes. They represent an additional layer of diversity to p97 complexes. Additional p97 interacting domains are the PUB [PNGase (peptide N-glycosidase)/ubiquitin-associated] (also known as PUG) domain of PNGase (Allen, Buchberger et al. 2006), the PUL domain of Doa1 (Mullally, Chernova et al. 2006). Several short p97-interacting motifs have been found, such as VIM (VCP-interacting motif) of various membrane-anchored proteins, such as the membrane-spanning ubiquitin E3 ligases (glycoprotein 78) (Ye, Shibata et al. 2004), (Zhong, Shen et al. 2004), SHP box of the yeast Derlin-1 homologue, Dfm1p (Sato and Hampton 2006) and the VBM (VCP-binding motif) of the polyglutamine-tract-containing protein, ataxin-3, a deubiquitinating enzyme (Boeddrich, Gaumer et al. 2006).

Additionally, many cofactors of p97 have ubiquitin interacting motifs and domains. Ufd1-Npl4 interacts with ubiquitin via a zinc finger domain at its C-terminus. p47 and contains an ubiquitin associating (UBA) domain at the N-terminus. p47 binds ubiquitin only when complexed with p97, and binds mono- rather than polyubiquitin conjugates. The UBA domain is required for the function of p47 in mitotic Golgi reassembly. Ubiquitin recognition seems to be a common feature of p97-mediated reactions.

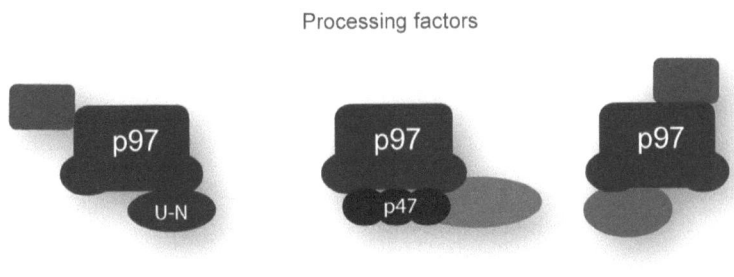

Fig. 5 p97 cofactor system. The cofactors direct p97 to specific cellular processes or support in the function of p97. p97 cofactors can be divided into two different classes namely adaptors / recruitment factors and processing factors. Several recruitments factors have been identified. For example, p47 and Ufd1-Npl4, which compete for binding to N-domain of p97 and thus form mutually exclusive complexes with the ATPase that target p97 for different pathways. Other recruitment factors recruit p97 complexes to membranes. Processing factors modify substrate proteins of p97, often by ubiquitination or deubiquitination. They bind mainly to other parts of p97 than the N-domain and therefore, do not compete with recruitment factors for binding to p97.

1.2.2 Roles of p97 in interphase processes

p97 function in Endoplasmic Reticulum associated degradation (ERAD)

Endoplasmic reticulum associated degradation

Proteins, which are targeted to the secretory pathway by a hydrophobic signal peptide at the N-terminus, are translocated into the Endoplasmic Reticulum (ER) lumen or integrated into the ER membrane through a channel (Johnson and van Waes 1999). Subsequently, they continue the export in ER-to-Golgi transport vesicles to the next compartment. It has been reported that an estimated 30% of all new synthesized proteins fail to achieve their native conformation and modifications (Schubert, Anton et al. 2000). This is dangerous for the cell since aggregation of misfolded proteins in the ER results in ER stress and induces the unfolded protein response. Saturation or failure of the unfolded protein response results in cell death or disease, including diabetes and late-onset neurological diseases (Kincaid and Cooper 2007). In the secretory pathway, high fidelity is achieved by a conserved enzymatic quality control system. Possibilities resulting in failing the quality control system are truncations, mutations, orphan subunits of hetero-oligomeric complexes, the lack of post-translational modifications (N-glycosylation, disulfide bond formation, GPI anchor addition or signal sequence cleavage) or misfolding. The ER resident quality control system governs the detection of defective proteins and subsequent rescue or retro-translocation into the cytosol for proteasomal degradation or deubiquitination (reviewed in (Trombetta and Parodi 2003)). The channel for retro-translocation channel is still not identified doubtlessly. ERAD has been investigated extensively and an important role for p97 has been found.

p97 function in endoplasmic reticulum associated degradation
An ERAD substrate, designated by the quality control system, is dedicated for retrotranslocation into the cytosol. A segment of the substrate is inserted into the ER membrane and retro-translocation is initiated. Next, polypeptides emerging into the cytosol are modified by lysine-48 polyubiquitination at the ER membrane by several ER-localized ubiquitin ligases. The complete retro-translocation into the cytosol is dependent on the function of p97 with the adaptor Ufd1-Npl4. $p97^{Ufd1-Npl4}$ complexes are recruited to the ER membrane and to ERAD substrates by integral membrane proteins containing p97 binding sites, like Ubx2, VIMP and membrane bound ligases. This recruitment requires the N-domain of p97. The ER membrane-bound receptors mediate assembly of the retrotranslocation machinery. Additionally, p97 recruits also deubiquitinating enzymes like the ataxin-3 to the ER membrane, but their function in the process of ERAD is not well understood (Wang, Li et al. 2006). The ubiquitin binding domain of Ufd1-Npl4 facilitates binding of the ubiquitinated substrate on the ER membrane. $p97^{Ufd1-Npl4}$ pulls the ubiquitinated substrate polypeptide out of the ER membrane using the energy generated by the its AAA-domains. Substrate binding of p97 at the ER membrane occurs only if a nucleotide is bound in the D1 ATPase domain of p97. ATP hydrolysis of the D2 domain is required for translocation of the substrate into the cytosol (Ye, Meyer et al. 2003). Subsequently, the ubiquitinated polypeptide chains are degraded by the proteasome in the cytosol (Figure 6) (reviewed in (Ye 2006)).

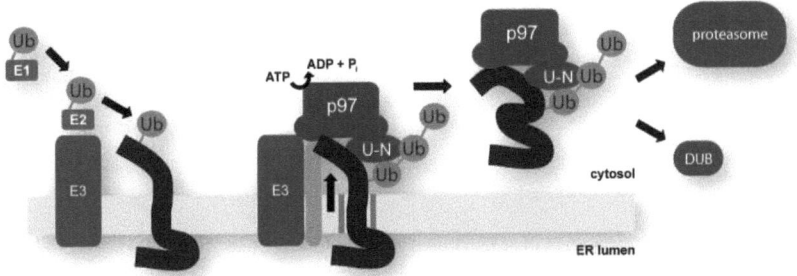

Fig. 6 Endoplasmic Reticulum associated degradation (ERAD). In the Endoplasmic Reticulum lumen resides a quality control system monitoring the correct maturation of new proteins targeted to the ER. An ERAD substrate, designated by the quality control system, is dedicated for retrotranslocation into the cytosol. Inserted in the ER membrane, it is ubiquitinated by E3 enzymes of the ERAD system. Subsequently, p97 associates and extracts the lysine-48 polyubiquitinated polypeptide out of the ER membrane in an energy dependent fashion. The polyubiquitinated polypeptides are degraded in the cytosol by the 26S Proteasome. Deubiquitinating enzymes are also recruited to this process by p97, but their role is not well understood.

p97 function in transcription activation

Involvement of p97 in activation of the transcription factors NF-κB, SPT23 and MGA2 has been shown. SPT23 and MGA2 are distant homologs of the p105 precursor of NF-κB.

The partially redundant transcription factors Spt23 and Mga2 of *Saccharomyces cerevisiae* play important roles in the regulation of the OLE1 and other lipid metabolism genes and control unsaturated fatty acid levels (Hoppe, Matuschewski et al. 2000), (Auld, Hitchcock et al. 2006). These ER membrane bound transcription factors are synthesized as 120 kDa precursors. Activation results in homo-dimerisation. One subunit of the homo-dimer undergoes ubiquitination at a specific site and a subsequent proteasomal processing resulting in an endoproteolytic cleavage event with Bidirectional restricted proteasomal degradation of the subunit. The amino terminal 90 kDa of the processed protein is not degraded, presumably due to the presence of a stable dimerisation domain. $p97^{Ufd1-Npl4}$ mobilizes the transcriptionally active 90 kDa fragment from the ER membrane bound partner anchor. The active 90 kDa fragment can enter the nucleus and activate transcription. Exact details of the p97-dependent mobilization are controversial in the different studies. One model suggests that the mono-ubiquitinated active 90 kDa fragment gets mobilized by $p97^{Ufd1-Npl4}$, whereas the 120 kDa precursor remains unmodified. (Rape, Hoppe et al. 2001). Recently a new model was presented, showing that $p97^{Ufd1-Npl4}$ binds to the polyubiquitinated 120 kDa fragment and promotes the segregation and mobilization of the unmodified active 90 kDa fragment (Shcherbik and Haines 2007) (Figure 7).

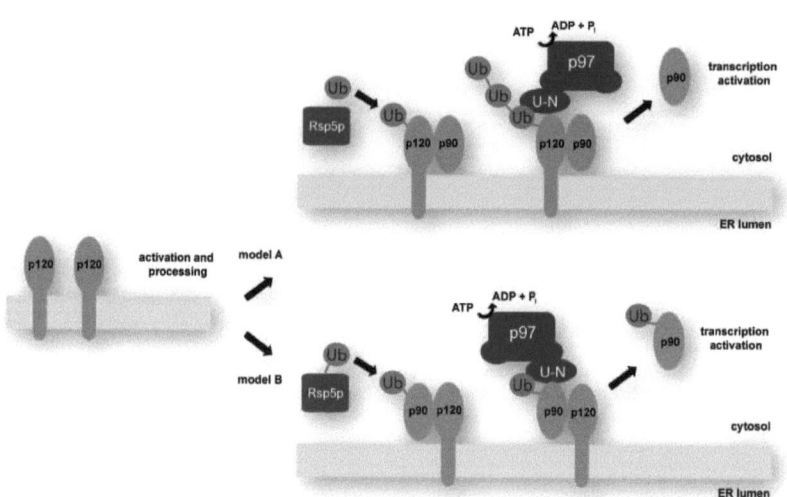

Fig. 7 Spt23/Mga2 mobilization by p97. The transcription factors Spt23/Mga2 of *Saccharomyces cerevisiae* play important roles in the regulation of the OLE1 and other lipid metabolism genes and control unsaturated fatty acid levels. These ER membrane-bound transcription factors are synthesized as 120 kDa precursors. Activation results in homodimerisation. One subunit of the homo-dimer undergoes ubiquitination and a subsequent proteasomal processing resulting in an endo-proteolytic cleavage event and bidirectional restricted proteasomal degradation. The amino terminal 90 kDa of the processed protein is not degraded. Two models exist that explain the p97-dependent step of the transcription activation. Model A illustrates that Spt23/Mga2p90 mono-ubiquitination serves as signal for its p97$^{Ufd1-Npl4}$ dependent mobilization. Model B proposes that p97$^{Ufd1-Npl4}$ promotes liberation of Spt23/Mga2p90 from the ER membrane via an interaction with poly-ubiquitinated Spt23/Mga2p120.

1.2.3 The roles of p97 in mitosis

In 1982, it has been found that a thermo sensitive mutant of CDC48, the yeast homologue of p97, arrests yeast cells in mitosis (Moir, Stewart et al. 1982). However, the reason for this arrest and the mechanism of p97 action in mitosis has remained elusive. Only now, the function of p97 in mitosis is starting to get unveiled, important mitotic processes have been found to be dependent on p97/CDC48. p97 activity is required for regulation of sister chromatid disjunction, spindle disassembly and formation of the nucleus, ER and Golgi late in mitosis.

The next section is dedicated to summarize the knowledge of p97 in mitosis. In all these processes the targets of p97 are not known, and therefore, the mechanisms are therefore unclear.

p97 function in Golgi reformation

The Golgi apparatus exists as a series of stacked flattened cisternal membranes. The Golgi is fragmented in mitosis, and reformed after mitosis. This process was studied in detail using an *in vitro* assay based on purified Golgi stacks, which are fragmented in mitotic cytosol. The well-established membrane fusion AAA ATPase NSF (N-ethylmaleimide-sensitive fusion protein) mediates the Golgi membrane fusion in a SNARE (soluble NSF-attachment protein receptor)-dependent manner (Rothman 1994). $p97^{p47}$ functions in a second pathway important for cisternal regrowth and shape. Golgi cisternae can be regrown in a cell-free system from mitotic Golgi fragments. Therefore, Golgi stacks are purified and incubated with mitotic cytosol and NEM (an alkylating agent inhibiting for example the AAA ATPases NSF and p97), which results a fragmented Golgi apparatus. Golgi fragments incubated with $p97^{p47}$ results in reformation cisternae, morphologically different from the NSF mediated regrowth (Kondo, Rabouille et al. 1997), (Shorter and Warren 1999). The combination of both pathways results in morphologically intact Golgi cisternae. The p97 adaptor protein p47 is required for this function. $p97^{p47}$ binds to syntaxin 5, a SNARE, and syntaxin 5 is regulated by $p97^{p47}$ (Rabouille, Kondo et al. 1998). This findings led the hypothesis that $p97^{p47}$, would be involved in a SNARE dependent fashion, similar to NSF. Phosphorylation of p47 by cdc2 in mitosis interferes with its Golgi membrane association and is required for Golgi cisternae disassembly during mitosis. p47 is localized mainly in the nucleus during interphase and released into the cytosol during mitosis (Uchiyama, Jokitalo et al. 2003). Additionally, p97 in collaboration with p37 is important for Golgi and ER maintenance in interphase (Uchiyama, Totsukawa et al. 2006).

In addition to that, the mitotic process requires ubiquitin and ubiquitin-binding by the UBA domain of p47, which is essential for Golgi reassembly at the end of mitosis and likely recruits the $p97^{p47}$ complex to a ubiquitinated substrate (Meyer, Wang et al. 2002). Additionally, the deubiquitinating activity of VCIP135, an OTU-type deubiquitinating cofactor of p97, is essential (Uchiyama, Jokitalo et al. 2002) (Wang, Satoh et al. 2004). VCIP135 is membrane bound and most likely deubiquitinates a substrate on mitotic Golgi fragments before reassembly can take place. The deubiquitinating step was shown to be a regulatory signal rather than a rescue from degradation, since proteasome inhibition did not abolish the requirement for

VCIP135 deubiquitinating activity. In contrast to the $p97^{p47}$ pathway, the $p97^{p37}$ pathway requires only the physical presence of VCIP135, but not its deubiquitinating activity (Uchiyama, Totsukawa et al. 2006). The targets of the $p97^{p47}$ and $p97^{p37}$ complexes and VCIP135 in reformation and maintenance of the Golgi/ER remain elusive, the SNARE syntaxin5 has been proposed, but it remains controversial since no direct evidences are present.

This represents processes of p97 that is independent of the proteasome but involving the ubiquitin system.

p97 function in sister chromatid disjunction

In fission yeast *Schizosaccharomyces pombe*, a connection between cdc48/p97 and Cut1/separase has been found. cdc48/p97 acts as a positive regulator of Cut1/separase and therefore of cohesin cleavage and sister chromatid disjunction. Destabilization of Cut1/separase caused by mutations of cdc48/p97 occurs in anaphase and takes place in a proteasome independent fashion. A direct stabilization of cut1/separase by p97 was suggested. It is not known whether the ubiquitin system is involved in this process and how the stabilization is achieved since no direct interaction between p97 and Cut1 was observed, but speculated to be during metaphase and anaphase (Ikai and Yanagida 2006).

p97 function in spindle disassembly

The mitotic spindle starts to disassemble after the chromatids have been segregated in telophase allowing chromosome decondensation and nucleus formation in the daughter cells at the end of mitosis. It was found that p97 with the adaptor Ufd1-Npl4 regulates spindle disassembly. In *Xenopus laevis* extract, absence of p97 function inhibits spindle disassembly. Microtubule dynamics, bundling and chromosome detachment are modulated by the $p97^{Ufd1-Npl4}$ complex. The underlying mechanism of the p97-dependent process is poorly understood. Data suggests that p97 regulates the interaction of spindle assembly factors like TPX2, Plx and XMAP215 with microtubules, but it is not shown, that this is the cause of the requirement of p97 in spindle disassembly (Cao, Nakajima et al. 2003). Recently, a report was published questioning the requirement of p97 in spindle disassembly (Heubes and Stemmann 2007).

p97 function in nucleus formation

The nuclear envelope (NE)

The nuclear envelope separates the cytoplasmatic from the nuclear compartment in eukaryotic cells. It is composed of two parallel lipid bilayers, the inner (INM) and the outer nuclear membrane (ONM). The ONM is continuous with the Endoplasmic Reticulum tubular network. The nuclear envelope is perforated by nuclear pore complexes. The nuclear pore complexes (NPCs) join the INM and the ONM and allow controlled the bidirectional movement of macromolecules between the nucleus and cytoplasm and NPCs are constituted of nucleoporin proteins bound to the membrane and linked to the chromatin and the lamina. The inner nuclear membrane is connected to the nuclear lamina, a fibrillar network consisting of intermediate filaments and inner nuclear membrane associated proteins (reviewed in (Margalit, Vlcek et al. 2005), (Gerace and Burke 1988), (Gant and Wilson 1997)) (Figure 8).

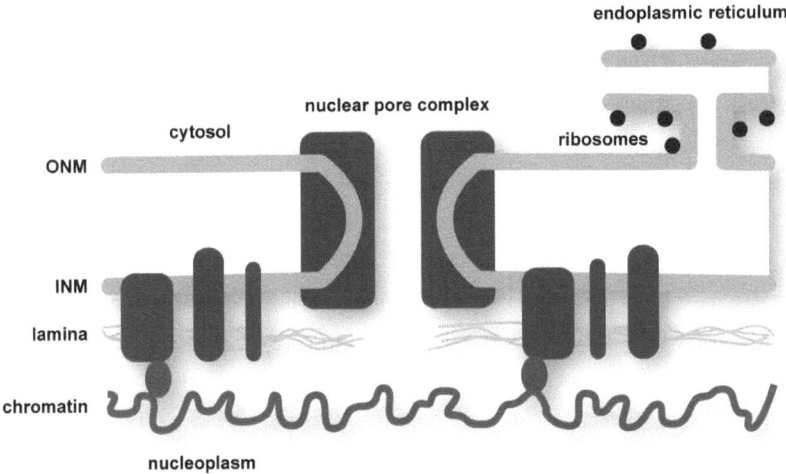

Fig. 8 Schematic representation of the Nuclear Envelope. Depicted is the nuclear envelope with the outer (ONM) and the inner nuclear membrane (INM) connected to the nuclear lamina and the chromatin, perforated by a nuclear pore complex.

Coordination of membrane fusion, lamina and NPC assembly during NE formation

During mitosis, the nuclear envelope breaks down in higher eukaryotes to allow the segregation of the replicated chromosomes. At the end of mitosis, the daughter nuclei reform around the segregated and decondensed chromatin. Late in anaphase, the reformation of the nucleus starts around the decondensing chromatin, thus requires coordination between the nuclear envelope fusion, NPC assembly and the nuclear lamina. The Nup107-Nup160 complex is the first nucleoporin complex that binds to chromatin (Walther, Alves et al. 2003). Membranes accumulate on chromatin accompanied by INM proteins LBR (lamin B receptor), LAP-1 and LAP-2 (Lamina-associated polypeptide) (Ellenberg, Siggia et al. 1997), (Yang, Guan et al. 1997) and early nucleoporins as the chromatin bound Nup153 and the membrane bound Pom121. Later the nucleoporins gp210 and Tpr associate with in the transition telophase to G_1-phase (Bodoor, Shaikh et al. 1999). The association of B-type lamins and A-type lamins to chromatin is detected in late anaphase or late telophase, respectively (Dechat, Gajewski et al. 2004).

Nuclear envelope membrane fusion is mediated by SNARE (soluble NSF attachment protein receptors) proteins in a NSF and alpha-SNAP (soluble NSF attachment factor) dependent fashion (Baur, Ramadan et al. 2007). The AAA ATPase NSF activates SNARS for fusion by disassembly of stable cis-SNARE complexes formed in previous membrane fusion reactions. NSF displays similar domain structure like p97. NSF is constituted of an N-domain and two AAA-domains, of with the first is active and required for SNARE complex disassembly and the second one required for hexamerisation. The NSF-N-domain is required for SNAP-SNARE complex binding (Nagiec, Bernstein et al. 1995), (Whiteheart, Rossnagel et al. 1994). Additionally, another mechanism for nuclear envelope formation was proposed. The ER and the nuclear envelope are connected, this led to the hypothesis that ER membrane tubules tips, from an ER network, attach to the decondensing chromatin and flatten to a nuclear envelope, rather than nuclear envelope formation starting from vesicular membranes by membrane fusion (Anderson and Hetzer 2007).

The coordination of NPC assembly and membrane fusion shows that chromatin decondensation and nuclear envelope formation is a process regulated a several stages. Two models exist in the literature for the coordination of nuclear membrane fusion and NPC assembly. The first model proposes a checkpoint mechanism based on the Nup107-Nup160 complex. Chromatin bound Nup107-Nup160 complexes sense the binding of membrane vesicles containing Pom121 and subsequently allow further membrane fusion and NPC assembly (Antonin, Franz et al. 2005). In contrast to this, the second model suggests NPC insertion into fused and flattened membrane cisternae. This model is based on experiments with blocked membrane fusion by NEM, an alkylating agent. (Macaulay and Forbes 1996). Newer finding support the second model. Specific inhibition of membrane fusion by excess of alpha-SNAP blocks NPC assembly at the stage in which Nup107 and gp210 are fully recruited to chromatin, but the assembly of FxFG repeat containing nucleoporins is blocked (Baur, Ramadan et al. 2007).

p97 involved in nucleus formation

It was found by investigation of the molecular requirements of nucleus formation in *Xenopus laevis* egg extract, that p97 was required for nucleus formation. The formation of a closed nuclear envelope around the chromatin requires the $p97^{Ufd1-Npl4}$ complex. Only patches and vesicles are detected on partially decondensed chromatin if $p97^{Ufd1-Npl4}$ is depleted. Subsequent nuclear growth is dependent on the $p97^{p47}$ complex (Hetzer, Meyer et al. 2001). The underlying molecular mechanisms of the requirement of p97 in nucleus formation remained unknown. The hypothesis of p97 being involved mechanistically in homotypic membrane fusion was a likely explanation as it was claimed for other processes, like the Golgi reformation after mitosis. But there exist differences. During the step, where extensive membrane fusion takes place, $p97^{Ufd1-Npl4}$ complexes are involved, unlike in the Golgi reassembly, where $p97^{p47}$ complexes are required. This could indicate a different mechanism of p97 function.

1.3 Functional interaction of p97 with Aurora B in *Xenopus laevis* egg extracts during nucleus formation

This thesis takes up the thread, where the previous research on the role of p97 in nucleus formation has stopped. This thesis presents the progress achieved in the understanding of the molecular mechanism of p97's role in nucleus formation. This project was performed in collaboration with a post doctoral fellow of our laboratory, Kristijan Ramadan. A summary of the part of the project done by Kristijan Ramadan in the laboratory follows next, whose work is closely connected to this thesis and will be presented as last part of the introduction of this thesis. The work of Kristijan Ramadan and the thesis writer Roland Bruderer cumulated in a publication of equal contribution (Ramadan, Bruderer et al. 2007).

For the understanding of the following part and the thesis as a whole it is important to introduce the *in vitro* nuclear envelope formation in *Xenopus laevis* egg extract in detail. Oocytes, eggs, egg extracts and embryos from the frog *Xenopus laevis* have been an important model system for studying cell-cycle regulation for several decades. The *Xenopus laevis* egg extract is an excellent *in vitro* system to study the process of nucleus formation on a molecular level.

1.3.1 Nuclear envelope formation in *Xenopus laevis* egg extract

Nuclear envelope formation can be recapitulated in an *in vitro* assay, which is based on *Xenopus laevis* egg extract. The lysis of *Xenopus laevis* eggs generates an extract containing the cytosolic components and membranes. This highly concentrated and active extract is capable of the decondensation of sperm chromatin and the formation of a closed nuclear envelope around the chromatin in the presence of energy (Lohka 1998). In a nucleus reformation assay, sperm chromatin is incubated in *Xenopus laevis* egg interphase extract. The sperm chromatin is hyper packed by sperm specific basic proteins X and Y. Decondensation of sperm chromatin is achieved by the acidic nuclear protein nucleoplasmin in an energy independent fashion. Nucleoplasmin replaces sperm specific basic proteins X and Y with egg histone H2A and H2B, resulting in assembly of somatic-type nucleosomes onto sperm DNA (Philpott and Leno 1992). In parallel to the decondensation, vesicles with a size of about 200 nm associate with the chromatin in an energy independent manner (Vigers and Lohka 1991) and (Newport and Dunphy 1992). Vesicle binding is followed by fusion and flattening of the membranes vesicles, which are energy dependent steps, forming membrane cisternae studded with NPCs covering the chromatin. In the next stage, a nucleus with a closed nuclear envelope forms around decondensed chromatin. Subsequently, the nuclei grow and the chromatin further decondenses, finally spanning about 7 µm in diameter. The formed nuclei are fully capable of transport and even full DNA replication (Figure 9).

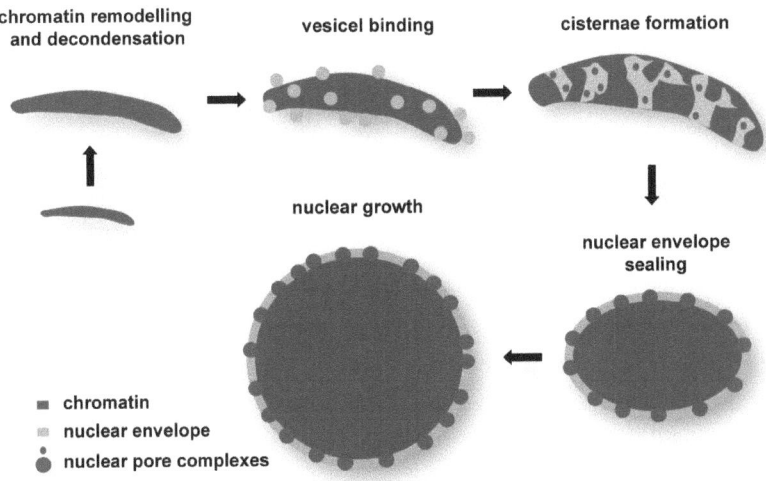

Fig. 9 Different steps during nucleus formation in *Xenopus laevis* egg extract. First, membrane vesicles bind to the chromatin. Binding is followed by fusion and flattening of the membrane vesicles. After final NE sealing, the nuclei are capable of growing.

1.3.2 Discovery of the p97$^{Ufd1-Npl4}$ Aurora B functional interaction

As described above, it was discovered that the p97$^{Ufd1-Npl4}$ complex is essential for nuclear envelope formation in *Xenopus laevis* egg extract (Hetzer, Meyer et al. 2001). The reason for the p97 requirement remains elusive from that work. Two hypotheses may explain the role of p97 in this process. The first hypothesis states that p97 may directly mediate membrane fusion, as it has been suggested for nuclear envelope formation and for Golgi / ER formation. An argument against this model is that SNARE dependent homotypic membrane fusion has been attributed to p97^{p47} complexes (Kondo, Rabouille et al. 1997) and not to the heterodimeric adaptor Ufd1-Npl4. According to the second model, p97 regulated nucleus formation by removing an inhibitor from the forming nucleus. This hypothesis is based on the analogy to the function of p97 in ERAD, where p97$^{Ufd1-Npl4}$ complexes extract polyubiquitinated ERAD substrate polypeptides out of the ER membrane.

Indications supported the second model suggesting that the p97-regulated inhibitor could be the mitotic chromosomal passenger complex kinase Aurora B. The first evidence was that p97$^{Ufd1-Npl4}$ physically interacts with the chromosomal passenger protein survivin in *Xenopus laevis* egg extracts (Vong, Cao et al. 2005). Second, Aurora B contributes to chromatin condensation during mitosis (Vagnarelli and Earnshaw 2004), (Hagstrom, Holmes et al. 2002), (Kaitna, Pasierbek et al. 2002), (Lipp, Hirota et al. 2007), (Takemoto, Murayama et al. 2007) and phosphorylates Histone H3 at serine 10, which regulates indirectly association of nuclear envelope membrane to chromatin (Hsu, Sun et al. 2000), (Fischle, Tseng et al. 2005), (Hirota, Lipp et al. 2005), (Kourmouli, Theodoropoulos et al. 2000). These processes may antagonize nucleus formation.

In order to test the second hypothesis, we set up a series of experiments in our laboratory. The first prerequisite of the hypothesis that p97 regulates an inhibitor of nucleus formation is that Aurora B actually is an inhibitor of nucleus formation. The addition of recombinant active wild-type (WT) Aurora B, but not a kinase-inactive form (K122R) nor the wt in combination with the Aurora B inhibitor hesperadin (Hauf, Cole et al. 2003) to *Xenopus laevis* egg extract prevented complete chromatin decondensation and nuclear envelope formation. This demonstrated that Aurora B activity is incompatible with nucleus formation. Importantly, the requirement for p97 in this process could be omitted by depletion or inactivation of Aurora B. Depletion of Aurora B from *Xenopus laevis* egg extract rendered the nucleus formation insensitive to addition of the dominant negative p97ΔD2 or addition of an inhibitory antibody against Ufd1. The same is also true if p97 is depleted and additionally the Aurora B kinase activity is inhibited by addition of hesperadin. Additionally, exit from mitosis (in Cytostatic factor arrested *Xenopus laevis* extracts by addition of Ca^{2+}) with mitotic chromatin showed that p97 inactivation results in retarded histone H3 dephosphorylation and chromatin decondensation. This shows that p97 activity antagonizes the kinase activity of Aurora B. Direct inactivation of Aurora B by addition of hesperadin results in immediate dephosphorylation and decondensation (Ramadan, Bruderer et al. 2007).

These findings were reproduced in *Caenorhabditis elegans* embryo *in vivo*. Two redundant homologues of p97 exist in *Caenorhabditis elegans*, CDC-48.1 and CDC-48.2. Depletion of CDC-48 again led to accumulation of condensed chromatin and inhibition of nuclear envelope formation. This shows that CDC-48 regulates nucleus formation as observed in the *Xenopus laevis* extract and that Aurora B regulation by p97 is conserved through evolution and species in the embryonic state. Additionally, depletion of CDC-48 again led to accumulation of condensed chromatin and inhibition of nuclear envelope formation, but also to an increase of embryos with phospho-H3-positive chromatin. Importantly, depletion of CDC-48 in the Aurora B temperature sensitive mutant air-2(or207) mutant worms at restrictive temperature restored H3 phosphorylation, whereas control RNAi directed against an unrelated gene adjacent to cdc-48.2 neither interfered with nucleus formation nor restored H3 phosphorylation. This notion was supported by our finding that feeding air-2(or207) mutant worms with bacteria expressing dsRNA to either cdc-48.1 or cdc-48.2 partially suppressed embryonic lethality (Ramadan, Bruderer et al. 2007).

These findings demonstrate that the mayor function of p97 in nucleus formation is to antagonize the kinase Aurora B. Thus, p97 has a regulatory role in nuclear envelope formation in *Xenopus* egg extract ruling out the possibility of p97 being involved directly in the mechanistic of homotypic membrane fusion events.

It remained unknown how and where p97 antagonizes Aurora B. To understand this newly discovered functional interaction, it was important to understand the molecular mechanism. Several fundamentally different ways are possible like the regulation of a phosphatase by p97, the regulation of Aurora B localization or the inhibition of the Aurora B kinase activity directly.

1.4 Aim of this thesis

As presented above, the role of $p97^{Ufd1-Npl4}$ in nucleus formation is to antagonize the mitotic kinase Aurora B. It is not understood, whether this functional interaction of $p97^{Ufd1-Npl4}$ with Aurora B is direct nor whether where and how it takes place.

The aim of this thesis was to shed light onto how $p97^{Ufd1-Npl4}$ complexes antagonize the mitotic kinase Aurora B during nucleus formation in *Xenopus laevis* egg extracts and understand the molecular mechanism.

Understanding the function of Aurora B and p97 is of great importance, since misregulation of these proteins often results in cell death or transformation.

2. Results

2.1 *In vitro* reconstitution of nuclear envelope formation and ubiquitination

To investigate the mechanism of p97 function in nucleus formation and especially NE formation, we chose the cell-free *Xenopus laevis* egg extract system that allowed us to study NE formation isolated from other processes like spindle disassembly and other mitotic events. This cell-free assay is based on *Xenopus laevis* egg interphase extract and demembranated sperm *Xenopus laevis* chromatin. It recapitulates nucleus and nucleus formation *in vitro* and can be easily manipulated biochemically. For a detailed description of what processes take place during chromatin decondensation and nuclear envelope formation see **1.3.1**.
Additionally, the *Xenopus laevis* egg extract system is also used to study earlier mitotic processes like chromatin replication, chromatin condensation and spindle assembly / disassembly. The *Xenopus laevis* egg extract can also be used to investigate cytosolic processes like for example ubiquitination and protein degradation or protein modification.

2.1.1 Preparation of crude and fractionated *Xenopus laevis* interphase extract

Xenopus laevis interphase egg extract was prepared from fresh *Xenopus laevis* unfertilized eggs. By injection of human chorionic gonadotropin, female *Xenopus laevis* frogs spawn eggs over night. The hormone induces transition of the eggs from metaphase I to metaphase II, in which eggs remain arrested due to the activity of a factor termed cytostatic factor (CSF) (Masui and Markert 1971). The freshly laid eggs have high cyclin B/cdc2 activity. Egg lysis by centrifugation causes Ca^{2+} influx from internal membrane stores (Han and Nuccitelli 1990), and influx from the extracellular space, which inactivates CSF and induces exit from mitosis. This is achieved by stimulating the degradation of cyclin B (Rauh, Schmidt et al. 2005). Addition of cycloheximide to the extract inhibits the synthesis of new proteins, most notably cyclin B and ensures that the extract remains in interphase. This is important for the experimental assay, since synthesis of new cyclin B would induce disassembly of newly formed nuclei followed by a subsequent re-entry into mitosis in the extract, interfering with analysis of nucleus formation in interphase.

Preparation of the extracts involves the following steps. For lysis, eggs are centrifuged at 37.500g for 20 min. This separates the crude interphase egg extract from lipids and yolk proteins. The crude interphase egg extract is used for the nuclear envelope formation assay. The crude interphase egg can be further fractionated. Subsequent fractionation of interphase egg extract is achieved by centrifugation of the crude extract at 200.000g for 70 min. During this fractionation step, cytosol is separated from light membranes, heavy membranes, ribosomes, and mitochondria (Figure 10a).

2.1.2 *In vitro* nucleus formation and quantification

The nuclear envelope formation reactions with the *Xenopus laevis* interphase egg extract and *Xenopus laevis* sperm chromatin were performed as follows. Demembranated sperm chromatin was incubated in crude interphase extract, produced as described above. Nucleus formation is an energy-dependent process (Marshall and Wilson 1997). Therefore, an ATP-RS (ATP-regeneration system) was added to supply all reactions with enough energy and to maintain constant ATP concentration during nucleus formation. The reactions were incubated at 19°C for different time periods. They were stopped by formaldehyde-fixation supplemented with the lipid dye 3,3'-dihexyloxacarbocyanine iodide ($DiOC_6$), to stain the membranes, and 4,6-diamidino-2-phenylindole (DAPI) to stain the DNA. Quantification of closed nuclear envelopes was performed by visual inspection of individual nuclei with epifluorescence microscopy. A nuclear envelope was considered closed, when the entire chromatin surface was covered with a smooth NE. 50 nuclei were analysed and counted randomly per reaction.

The *Xenopus laevis interphase* egg extracts used in this thesis if it satisfied the following criteria. If decondensed sperm chromatin, completely enclosed by a NE, was observed after a time window of 45 to 90 min, the extract was used.

The progress of formation of closed nuclear envelopes around decondensed sperm chromatin was monitored over time. Aliquots were taken and fixed at multiple time points. Figure 10b shows representative intermediates. The sperm chromatin starts to decondense in the first 10 min, additionally membrane vesicles start to bind to the decondensed sperm chromatin. Patches of fused and flattened membrane vesicles forming membrane cisternae are visible on the chromatin surface after 20 min. Subsequently, the chromatin decondenses further and membrane cisternae grew and covered chromatin surface. Most of the sperm chromatin particles rounded up and had a closed nuclear envelope after 60 min (Figure 10b).

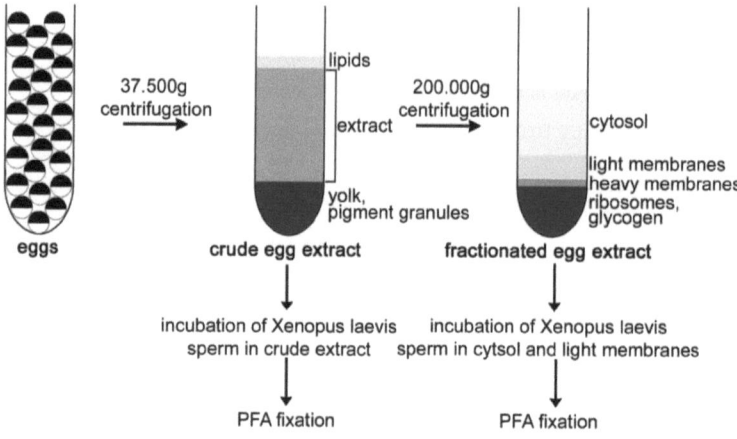

Fig. 10a Schematic overview on preparation of crude and fractionated *Xenopus* egg extract.

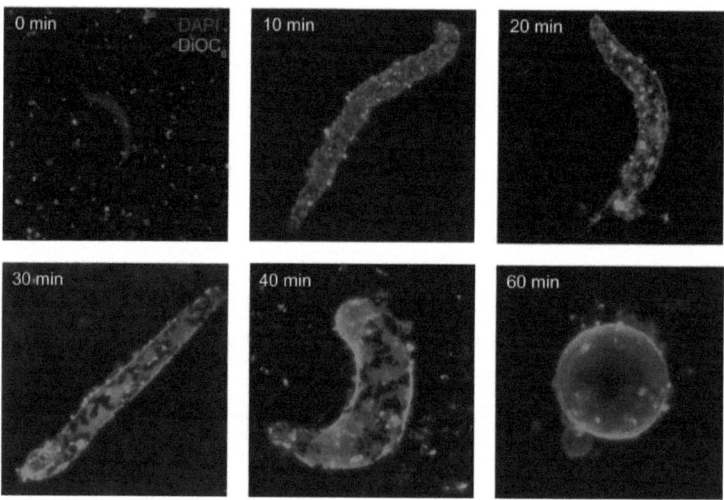

Fig. 10b Time course of nuclear envelope reformation. Nuclear envelopes were formed by incubating Xenopus laevis egg cytosol and membranes with demembranated sperm chromatin. Samples were fixed at indicated times. Chromatin and membranes were stained with 4,6-diamidino-2-phenylindole (DAPI) and 3,3-dihexyloxacarbocyanine iodide (DiOC6), respectively, and imaged by confocal fluorescence microscopy. Shown are maximum intensity projections of confocal sections.

2. Results

2.1.3 *In vitro* ubiquitination analysis using *Xenopus laevis interphase* egg cytosol

These *Xenopus laevis interphase* egg extracts, capable of the recapitulation of the cellular processes of nucleus formation and endoplasmic reticulum formation, were used to analyse the ubiquitination of proteins of interest. These extracts mimic a situation during exit of mitosis. The ubiquitination length and chain-types of proteins of interest could be analysed. The ubiquitination reactions were performed with addition of exogenous HA-tagged ubiquitin to the extracts, which allows very sensitive detection of ubiquitin modified proteins. Since ubiquitinated proteins are often in complex with additional factors denaturing immunoprecipitations were performed.

Xenopus laevis interphase egg cytosol was incubated with recombinant HA-tagged ubiquitin wt or mutants containing ATP-RS. Energy was added since ubiquitination is an energy dependent process. The reactions were incubated for 30 min at 19°C. Next, the reactions were stopped by either denaturation in SDS or addition of N-ethylmaleimide (NEM) that alkylates cysteine proteases and thus prevents deubiquitination and/or degradation during subsequent analysis. The NEM treatment of the reaction was important since ubiquitinated proteins are not stable. Either they are degraded by the proteasome system or polyubiquitin chains are reduced and eliminated by deubiquitinating enzymes, which reside in the cytosol and do not require energy for catalysis. Subsequently, immunoprecipitations for the proteins of interest were performed and the status of ubiquitination was analysed by denaturing gel electrophoresis and western blotting.

2.2. The first evidence for a role of the chromosomal passenger complex in nucleus formation

This thesis takes up the thread, where the previous research on the role of p97 in nucleus formation has stopped. From the previous work, it was known that the activity of the AAA-ATPase p97 is required for successful spindle disassembly and nucleus formation during the exit of mitosis. This function of p97 also requires the heterodimeric recruiting factor Ufd1-Npl4. The molecular mechanism of the requirement of $p97^{Ufd1-Npl4}$ remains elusive from that work (Hetzer, Meyer et al. 2001).

We set out to understand the requirement of p97 in nucleus formation after mitosis. The role of $p97^{Ufd1-Npl4}$ in this process can be explained by at least two hypotheses. A first hypothesis states that p97 directly mediates membrane fusion. This has been suggested for nuclear envelope formation and Golgi/ER reformation. An argument against this model is that SNARE dependent homotypic membrane fusion has been attributed to $p97^{p47}$ complexes (Kondo, Rabouille et al. 1997) and not to the heterodimeric recruiting factor Ufd1-Npl4. According to a second model, p97 regulated nucleus formation by removing an inhibitor from the forming nucleus. This hypothesis is based on the analogy to the function of p97 in ERAD, where $p97^{Ufd1-Npl4}$ complexes extract polyubiquitinated ERAD substrate polypeptides out of the ER membrane. We favored the second model and chose a hypothesis driven approach. An interesting interaction was discovered that drew our attention. It was found that survivin interacts with p97 and Ufd1-Npl4 in *Xenopus laevis* egg interphase cytosol (Vong, Cao et al. 2005). Survivin is a member of the chromosomal passenger complex, which also contains the kinase Aurora B, INCENP, Dasra, and TD60. The chromosomal passenger complex is active in mitosis, it first localizes to chromatin, then to kinetochores and next to the midbody. Its active component is the kinase Aurora B, which phosphorylates a variety of chromatin bound proteins and regulates chromatin associated processes as well as cytokinesis.

Since the chromosomal passenger complex is reported to be involved in a variety of mitotic processes like chromatin condensation, spindle formation and stability or chromatin segregation (see 1.1.2 for a detailed description); we hypothesized that the chromosomal passenger complex could have an inhibitory role in nucleus formation late in mitosis. It is known that Aurora B contributes to chromatin condensation during mitosis (Vagnarelli and Earnshaw 2004), (Hagstrom, Holmes et al. 2002), (Kaitna, Pasierbek et al. 2002), (Lipp, Hirota et al. 2007), (Takemoto, Murayama et al. 2007) and phosphorylates Histone H3 at serine 10, which regulates indirectly association of nuclear envelope membrane to chromatin (Hsu, Sun et al. 2000), (Fischle, Tseng et al. 2005), (Hirota, Lipp et al. 2005) and (Kourmouli, Theodoropoulos et al. 2000). These processes regulated by the chromosomal passenger complex via Aurora B may antagonize nucleus formation.

The finding that p97 and survivin interact and the known functional mechanism of $p97^{Ufd1-Npl4}$ in the process of ERAD led to the hypothesis, that p97 could negatively regulate survivin activity. Therefore, we started this project by monitoring, how enhanced levels of survivin effect nucleus formation. We performed this by addition of exogenous survivin to *Xenopus laevis* egg interphase extract. Survivin was reported to activate the kinase Aurora B (Bolton, Lan et al. 2002). Thus, we reasoned that a phospho-mimic version of survivin, which interferes with the interaction of survivin with the chromosomal passenger complex, could serve as a negative control to the addition of exogenous survivin wt.

Survivin excess inhibits NE formation in Xenopus laevis egg extracts

The above stated rational led us to set up the following. We analysed whether addition of exogenous recombinant survivin could inhibit nucleus formation. We received a bacterial expression construct containing a survivin fusion protein with an N-terminal Glutation-S-Transferase (GST) tag from Zheng, Y (Vong, Cao et al. 2005). Interestingly, addition of this survivin fusion protein containing GST tag inhibited nucleus formation in *Xenopus laevis* interphase egg extract reactions. The GST is a large, dimeric tag and close to the N-terminal dimerisation domain of survivin (Figure 11a), concluded from the crystal structure (Chantalat, Skoufias et al. 2000). Therefore the GST tag could interfere with dimerisation and function of survivin. Therefore a new His-tagged construct was cloned. The generation of His-tagged survivin did not result in soluble protein, possibly because survivin has a zinc binding BIR domain at the N-terminus.

Hence, we generated a new bacterial expression construct of survivin with the small C-terminal *strep*-tag. The *strep*-tag is a short peptide (8 amino acids) with highly selective binding properties for a streptavidin variant which has been named "*strep*-tactin". Elution is very mild and performed by addition of 10 mM biotin. As negative control for the addition of exogenous *strep*-tagged survivin wt in nucleus formation reactions, we constructed the phosphomimic version of survivin containing a point mutation of serine 126 to a glutamate at the putative phosphorylation site of Aurora B (S126E). Phosphorylation at this site was reported to interfere with survivin localization in somatic cells. Additionally, it interferes with interactions of survivin with the rest of the chromosomal passenger complex (Wheatley, Henzing et al. 2004), mostly INCENP and Borealin (Figure 11b), concluded from the structure of the complex (Jeyaprakash, Klein et al. 2007).

Recombinant *strep*-tagged survivin wt and survivin S126E were expressed in *E. coli*. First step of the purification was a streptactin affinity chromatography. Further purification of the streptactin purification eluate was performed by gelfiltration using a preparative Superdex 200 column. Survivin wt and S126E eluted with a sharp peak at volume corresponding to an about 80 kDa protein (Figure 11c and 11d) as judged using a gelfiltration calibration kit. A small peak at the volume corresponding to the void could be separated, this peak contained also a small fraction of aggregated survivin. The recombinant survivin eluted at a higher molecular weight than expected for a dimeric molecule of 36 kDa weight. This may be explained by its elongated shape (Figure 11a), since gelfiltration chromatography elution properties of a protein depend on relative size of the protein molecules (Deutscher, Abelson et al. 1990). Subsequently, the gelfiltration peak fractions were further purified using MonoQ anion exchange chromatography and eluted with a linear gradient of 50 mM to 1000 mM KCl. Survivin wt and S126E eluted both at a salt concentration of 440 mM KCl. The purifications were monitored by SDS PAGE (Figure 11c and 11d). This purification procedure resulted in highly pure proteins. The resulting fractions were concentrated using an Amicon Ultra 10'000 MWCO filtration device, desalted in ELB (Egg lysis buffer) and stored after snap freezing in liquid nitrogen at -80°C in small aliquots.

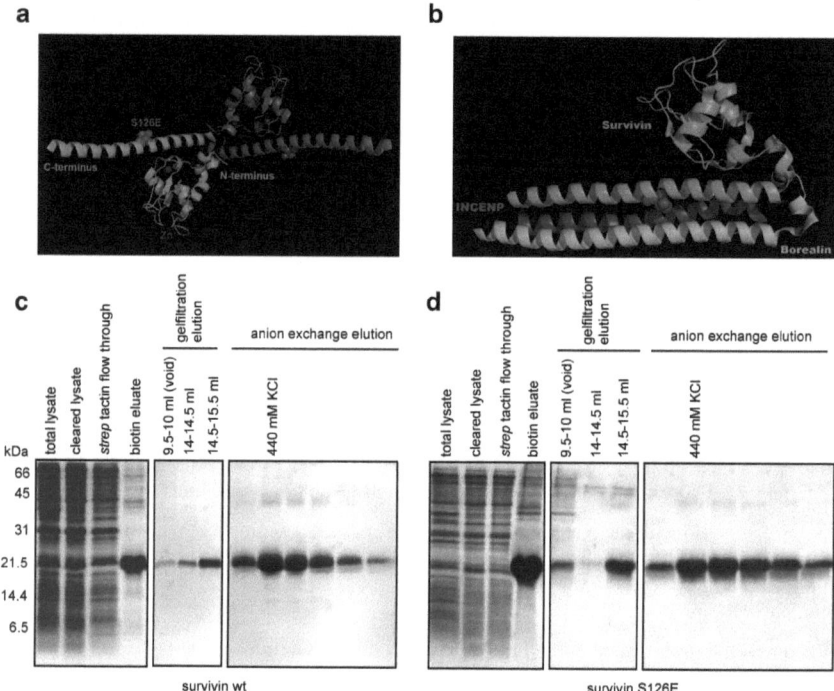

Fig. 11 Purification of survivin wt and survivin S126E. a, structure of a survivin dimer with the site of mutation coloured in red (Chantalat, Skoufias et al. 2000). **b**, structure showing the interaction of survivin with INCENP via the C-terminal helix (Jeyaprakash, Klein et al. 2007). **c**, *strep*-tagged survivin wt was expressed in *E. coli* and purified via *strep* tactin affinity chromatography, gelfiltration with a Superdex 200 column and MonoQ anion exchange chromatography. Analysis of the purification by Coomassie stained SDS PAGE gels of the *strep* tactin affinity purification, gelfiltration and anion exchange chromatography. survivin wt eluted at 440 mM KCl. **d**, *strep*-tagged survivin S126E was expressed and purified as in A. survivin S126E eluted at 440 mM KCl.
Note the weak band at approximately 40 kDa is survivin (most likely dimeric form) as analysed by western blotting.

The purified *strep*-tagged survivin wt and S126E preparations were added to nucleus formation reactions *Xenopus laevis* interphase egg extract at different concentrations ranging from 5 to 27 μM survivin wt or survivin S126E. The reactions were fixed after 90 min and imaged by epifluorescence microscopy. Interestingly, survivin wt had a strong effect on chromatin decondensation and nuclear envelope formation, whereas survivin S126E had no inhibitory effect (Figure 12a). Quantification of the reactions revealed a concentration dependent inhibition of survivin wt and virtually no detectable inhibition of survivin S126E addition in the concentrations tested (Figure 12b). Since the phospho-mimic survivin S126E has no effect, this suggests an active role of Aurora B in this process. We focused in the following research on the kinase Aurora B. Motivated by these findings, Kristijan Ramadan tested directly the influence of the kinase activity of Aurora B on nucleus formation. He could show that elevated levels of Aurora B by addition of exogenous, active Aurora B is incompatible with nucleus formation. This demonstrates that Aurora B is an inhibitor of nucleus formation. Importantly, the requirement for p97 in the process of nucleus formation could be omitted by depletion or inactivation of Aurora B. This shows that p97 antagonizes Aurora B. It was not understood, if this functional interaction is direct, nor was it clear where p97 antagonizes Aurora B. To address the molecular mechanism of this functional interaction we started with the investigation of the physical interaction between Aurora B and p97.

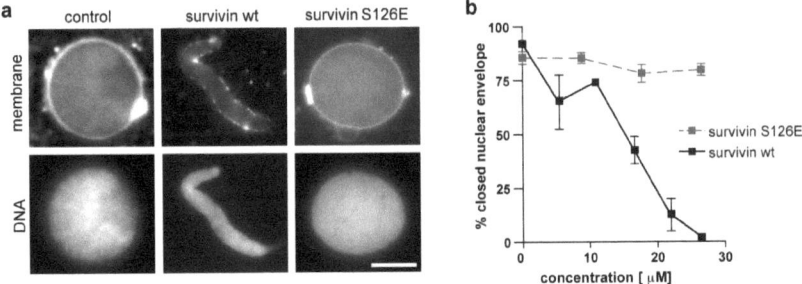

Fig. 12 Exogenous survivin wt but not survivin S126E inhibits nuclear envelope formation. a, nuclear envelopes were formed by incubating *Xenopus laevis* egg interphase extract with demembranated sperm chromatin. Reactions were carried out with increasing concentrations of survivin wt or survivin S126E (purified in Fig. 3), or survivin buffer as control. After 90 min, the samples were fixed and imaged by epifluorescence microscopy. Scale bar, 5 μm. **b,** the percentage of chromatin particles with closed nuclear envelopes was determined visually by light microscopy in reactions performed as in (a) at the indicated concentrations of survivin wt or S126E. Shown are means from 5 independent experiments (n=5) with >50 particles counted in each (error bars show STDEV.)

2.3 Interaction analysis of p97 with the chromosomal passenger complex

Next we characterized the interaction of p97 with the chromosomal passenger complex in detail. Therefore, we successfully generated new polyclonal antibodies against the *Xenopus laevis* proteins GST-survivin, GST-Aurora B and against a C-terminal peptide of INCENP. Using these new antibodies, we immunoprecipitated the chromosomal passenger complex from *Xenopus laevis* egg interphase cytosol under native conditions. Western blotting analysis revealed that all antibodies immunoprecipitated quantitatively the corresponding proteins, used as antigens (Figure 13). Probing the western blot membranes for the Aurora B, survivin and INCENP revealed that in all immunoprecipitations, the other chromosomal passenger complex proteins co-immunoprecipitated. This result was consistent with published data (Bolton, Lan et al. 2002). Subsequently, we probed for co-immunoprecipitation of p97. We could confirm the binding of survivin to p97 (Vong, Cao et al. 2005). Interestingly, immunoprecipitation of Aurora B also co-isolated p97. This is interesting, since the kinase Aurora B is the active part of the chromosomal passenger complex. We could not detect any specific co-immunoprecipitation of p97 in INCENP immunoprecipitations.

This shows that complexes between survivin and p97 as well as between Aurora B and p97 are present in cytosol. It is not clarified, if they are separate complexes or a ternary complexes. But importantly, it shows a interaction of the kinase Aurora B with p97 does occur.

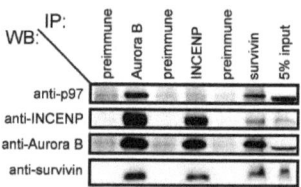

Fig. 13 Interaction of Aurora B and survivin but not INCENP with p97. Aurora B, INCENP and survivin were immunoprecipitated (IP) from *Xenopus laevis* egg interphase cytosol under native conditions using sera and corresponding preimmune sera. The isolates were analysed by western blotting (WB). Note that INCENP IP co-immunoprecipitated Aurora B and survivin but not p97.

p97 is an ubiquitin dependent chaperone. In other processes like the processing for the transcription factors or the ERAD, p97 interacts with proteins after their ubiquitination. We first asked whether Aurora B and survivin are ubiquitinated and what type of ubiquitin modification it is present. Next, we addressed if p97 interacts with the chromosomal passenger complex proteins Aurora B and survivin after their modification by ubiquitination.

We set up an assay to isolate and monitor ubiquitinated proteins in *Xenopus laevis* egg interphase cytosol as it is explained in 2.1.3. We generated recombinant HA-tagged ubiquitin wt, K46R and K63R. The K48R variant contains an asparagine instead of lysine at position 48, the chain extension at lysine-48 is the principal modification for proteasomal degradation (reviewed in (Pickart 1997)). The chain extension on lysine-63 is reported to be a regulatory ubiquitination (reviewed in (Haglund and Dikic 2005)). Recombinant HA-tagged Ubiquitin wt, K46R and K63R were expressed in bacteria, purified by acid precipitation of bacterial proteins with 3.5% perchloric acid, dilution and subsequent ion exchange chromatography was performed using a Hitrap SP column. Finally, the proteins were dialysed and concentrated with an Amicon Ultra filtration device 5000 MWCO (Figure 14a). We incubated the HA-tagged ubiquitin variants at 20 µM in *Xenopus laevis* egg interphase cytosol with energy, to test if the recombinant HA-tagged ubiquitin variants could be used in the cytosol for modification of proteins. Western blot analysis revealed that indeed the recombinant protein was incorporated in polyubiquitin chains. The HA-tagged ubiquitin wt and the K63R displayed approximately the same degree of incorporation in chains. The variant K48R incorporated to a lower extent into chains (Figure 14b).

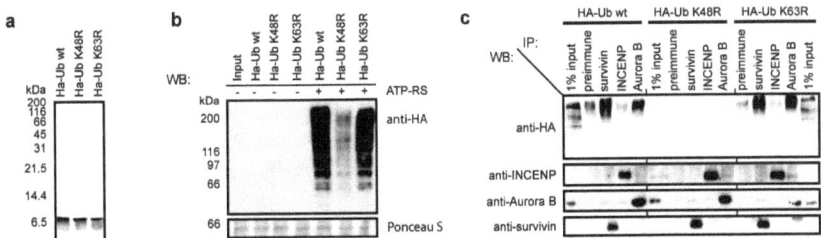

Fig. 14 Aurora B and survivin are polyubiquitinated. a, Analysis of the purity of the HA-tagged ubiquitin wt, K48R and K63R by SDS PAGE stained with Coomassie. **b**, *Xenopus laevis* egg interphase cytosol was incubated with 20 µM HA-tagged wt ubiquitin or the chain extension mutants K48R or K63R for 30 min. Polyubiquitination was detected by western blotting (WB) with HA-specific antibodies. **c**, *Xenopus laevis* egg interphase cytosol was incubated for 30 min with 20 µM HA-tagged wt ubiquitin or the chain extension mutants K48R or K63R, and denatured in 1% SDS to break protein complexes. After renaturation, Aurora B, survivin and INCENP were immunoprecipitated (IP) and analysed, on a gradient gel ranging from 7.5% to 18% SDS by western blotting.

We then asked whether p97 interacts with the non-modified or the ubiquitinated forms of Aurora B and survivin. Therefore, we analysed the ubiquitination state of, Aurora B, survivin and INCENP in cytosol. We incubated *Xenopus laevis* egg interphase cytosol with 20 μM HA-tagged ubiquitin variants for 30 min. Next, we denatured the proteins in the reactions with 1% SDS to break the complexes. Then, we diluted to reactions to 0.1% SDS with Phosphate Buffered Saline (PBS) containing 1% Triton X-100 and 1% Bovine serum albumin to renature and allow antibody binding. We preformed immunoprecipitations of Aurora B, survivin and INCENP. The precipitates were separated on a gradient gel. Western blot analysis revealed that Aurora B and survivin were modified in cytosol by lysine-48 linked ubiquitin chains. No modification was detected for INCENP (Figure 14c).

This demonstrates that ubiquitinated Aurora B and survivin are present in cytosol modified by lysine-48 chains. Hence, we asked whether p97 binds the modified forms.

Next, we investigated if p97 interacts with the ubiquitinated forms of Aurora B. We analysed the interaction in two ways, by pull down of His-tagged p97 or by immunoprecipitation of endogenous p97$^{Ufd1-Npl4}$ complexes with a monoclonal Ufd1 antibody.

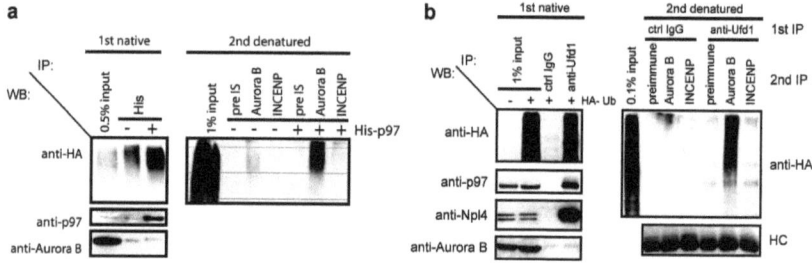

Fig. 15 Interaction of p97$^{Ufd1-Npl4}$ with polyubiquitinated Aurora B. a, *Xenopus laevis* egg interphase cytosol was incubated with 95 μM HA-ubiquitin with or without 0.02 μg/μl His-tagged p97 for 30 min and subjected to anti-RGS-His pull down. One half was directly analysed by western blotting (WB) (first IP native). The other half was denatured, split into three and subjected to immunoprecipitation with indicated antibodies (second IP denatured) and subsequently analysed by western blotting. **b**, *Xenopus laevis* egg interphase cytosol was incubated with 95 μM HA-ubiquitin for 30 min and subjected to anti-Ufd1 or control immunoprecipitation. One half was directly analysed by western blotting (first IP native). The other half was denatured, split into three and subjected to immunoprecipitation with indicated antibodies (second IP denatured) and subsequently analysed by western blotting. Heavy chains (HC) are Ponceau S staining. Note that only ubiquitinated Aurora B co-isolated with p97$^{Ufd1-Npl4}$.

For the first method, we incubated *Xenopus laevis* egg interphase cytosol with at 95 µM HA-tagged ubiquitin wt with or without 0.02 µg/µl exogenous His-tagged p97 for 30 min. Then, the reactions were stopped by cooling on ice and addition of 5 mM N-ethylmaleimide to prevent deubiquitination. Subsequently, His-p97 was isolated under native conditions with monoclonal anti-RGS-His antibodies. One half of the reactions (1st native) was directly analysed by western blotting (Figure 15a, left panel). This revealed that ubiquitinated proteins co-isolated with p97, and analysis for unmodified Aurora B showed no specific co-isolation over background level. The other half of the reactions (2nd denatured) was denatured with 1% SDS to break the complexes. Then, we diluted to reactions to 0.1% SDS with Phosphate Buffered Saline (PBS) containing 1% Triton X-100 and 1% Bovine serum albumin to renature and allow antibody binding. The resulting renatured reactions were split in three and subjected immunoprecipitations with control, Aurora B or INCENP antibodies (Figure 15a, right panel).

Interestingly, His-p97 binds polyubiquitinated Aurora B but not the unmodified form in cytosol. To further establish this important finding, the interaction of polyubiquitinated Aurora B to endogenous p97 complexes was analysed.

The second approach to analyse the interaction of p97 with ubiquitin modified Aurora B had the advantage that it was done without addition of exogenous p97. *Xenopus laevis* egg interphase cytosol was incubated with 95 µM HA-tagged ubiquitin wt for 30 min. Then, the reactions were stopped by cooling on ice and addition of 5 mM N-ethylmaleimide to prevent deubiquitination. Subsequently, $p97^{Ufd1-Npl4}$ complexes were immunoprecipitated with the anti-Ufd1 monoclonal antibody 5E2 under native conditions. One half of the reactions (1st native) was directly analysed by western blotting (Figure 15b, left panel). This revealed that $p97^{Ufd1-Npl4}$ co-isolates ubiquitinated proteins, and analysis for unmodified Aurora B showed no specific co-isolation over background. The other half of the reactions (2nd denatured) was subjected to denaturing immunoprecipitations with control, Aurora B or INCENP antibodies as described above (Figure 15b, right panel). This showed that $p97^{Ufd1-Npl4}$ interacts with ubiquitinated, but not the unmodified form of Aurora B.

This confirmed that also endogenous $p97^{Ufd1-Npl4}$ complexes bind only polyubiquitinated forms of Aurora B but not the unmodified form. This is a strong indication that Aurora B could be a substrate of p97. Therefore, we asked whether the putative substrate Aurora B is regulated by p97.

2.4 Inactivation of p97 results in accumulation of ubiquitinated Aurora B and survivin

The previous experiments show, that p97 interacts with ubiquitinated forms for Aurora B and hence Aurora B could be a substrate of p97. p97 acts downstream of ubiquitination and hands over substrates to the proteasome or to deubiquitination enzyme. Thus, we asked if p97 regulated the amount of polyubiquitinated Aurora B in cytosol.

To be able to investigate the effect of p97 on Aurora B, we inactivated p97 in the *Xenopus laevis* egg interphase cytosol. To interfere with the function of p97 two different methods were applied, the first is a dominant-negative approach and the second approach is depleting p97.

The dominant-negative approach takes advantage of the use of a dominant-negative fragment of p97. This fragment does not contain the second AAA domain (Figure 4), which is the ATPase active AAA domain of p97 (Ye, Meyer et al. 2003). It is termed p97ΔD2. Since ATP binding in the first AAA module of p97 is required for substrate binding a mutation in the Walker A motive of the first AAA module (K251A) renders p97ΔD2 from at dominant negative to a negative. Therefore, p97ΔD2 wt interferes with endogenous p97 activity and p97ΔD2 K251A serves as a control since it has no effect (Baur, Ramadan et al. 2007). Both proteins were expressed in *E.coli*. First, affinity purification step Ni^{2+} affinity chromatography was performed, followed by gelfiltration using a Sephacryl 300 column and ion-exchange chromatography using a MonoQ column, the protein eluted at 350 mM KCl. This purification resulted in highly pure proteins (Figure 16).

For the dominant negative inhibition of p97 approach, *Xenopus laevis* egg interphase cytosol was incubated with 20 μM HA-tagged ubiquitin wt containing 8 μM hexamer p97ΔD2 wt or storage buffer for 30 min. Subsequently, the reactions were subjected to denaturing immunoprecipitations with control, Aurora B, survivin or INCENP antibodies as described above (Figure 17a).

The levels of polyubiquitinated Aurora B and survivin increases upon inactivation of p97 function with p97ΔD2. Polyubiquitinated INCENP did not accumulate in absence of p97 activity.

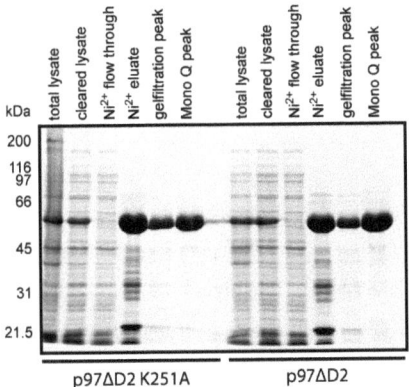

Fig. 16 Purification of p97ΔD2 and p97ΔD2 K251A. Bacterially expressed His-tagged p97ΔD2 and p97ΔD2 K251A were first purified by Ni^{2+} affinity chromatography, followed by gelfiltration using a Sephacryl 300 column and ion-exchange chromatography using a MonoQ column. Purity after the different purification steps was analyzed by SDS-PAGE stained with Coomassie.

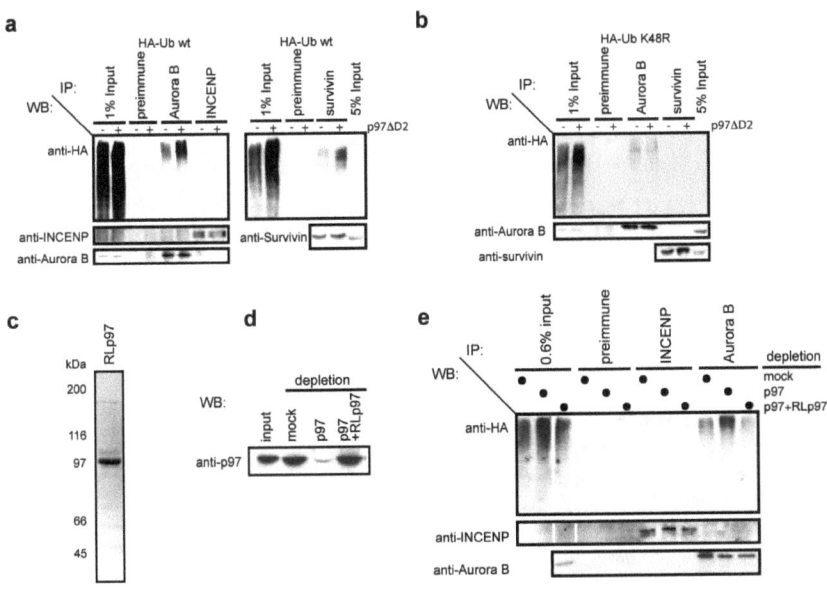

Fig. 17 p97 influences cytosolic K48-polyubiquitylation of Aurora B and survivin. a, *Xenopus laevis* egg interphase cytosol was incubated with 20 μM HA-tagged wt ubiquitin and 8 μM p97ΔD2 or p97ΔD2 buffer for 30 min. Subsequently, the reactions were subjected to denaturing immunoprecipitation of Aurora B, INCENP or preimmune serum as control. The efficiency of Aurora B, survivin and INCENP isolation was controlled with specific antibodies. **b,** the reactions were performed as in (a)

but 20 μM HA-tagged ubiquitin K48R was added. **c**, p97 was purified from rat liver tissue. Analysis of the purification by SDS PAGE stained with Coomassie. **d**, *Xenopus laevis* egg interphase cytosol was mock-depleted, p97-depleted or p97-depleted and reconstituted with addition of p97 (purified from rat liver (c)). Analysis performed by western blotting (WB). **e**, the cytosols from (d) were supplemented with 20 μM HA-tagged wt ubiquitin and incubated for 30 min, denatured and subjected to immunoprecipitation of Aurora B, INCENP or preimmune serum as control. Polyubiquitination was detected by western blotting with HA-specific antibodies.

We modified and repeated the experiment above to analyse, which kind of ubiquitin chains accumulate in absence of p97 function. Instead of 20μM HA-tagged ubiquitin wt the K48R variant was added to the reactions. The signal for cytosolic polyubiquitination was much weaker than in the experiment with HA-tagged ubiquitin wt. The western blot membranes were exposed longer to the film. This did not result in increase in polyubiquitin chains of HA-tagged ubiquitin K48R detected upon inactivation of p97 (Figure 17b).

Together these experiments show that the absence of p97 function results in accumulation of lysine-48 chain ubiquitin modification containing Aurora B and survivin in cytosol. To further verify this, the second method for inhibiting p97 activity was performed.

The second approach is by depletion of p97 complexes. p97 can be depleted from *Xenopus laevis* egg interphase cytosol using a fragment of Ufd1 termed UT6 (comprising residues 215-207), that binds to p97 and specifically depletes p97 but not the adaptors Ufd1-Npl4 or p47 (Hetzer, Meyer et al. 2001). Restoration of p97 in the depleted reactions was performed using p97 purified from rat liver. RLp97 was purified from rat liver cytosol. The first step was a 30% ammonium sulphate precipitation, followed by anion exchange chromatography using a Hitrap Q column, p97 eluted at about 380 mM KCl. Next, gelfiltration with a Superdex200 column was preformed (Figure 17c).

In this approach, we analysed the regulation of polyubiquitination of Aurora B using depletion of p97. We depleted *Xenopus laevis* egg interphase cytosol mock or of p97 complexes with GST or GST-UT6, respectively. This resulted in a reduction of p97 levels to about 1% in the depleted cytosol. Additionally, we reconstituted p97 depleted cytosol with 0.05 μg/μl p97 purified from rat liver (Figure 17d). Next the reactions were subjected to denaturing immunoprecipitations with control, Aurora B or INCENP antibodies as described above. The levels of polyubiquitinated Aurora B increased upon depletion of p97. INCENP was not affected. Reconstitution of the cytosol resulted in levels of polyubiquitinated Aurora B close to mock depletion (Figure 17e).

The observation that the levels of polyubiquitinated Aurora B in cytosol containing lysine-48 chains are regulated by p97 raises the possibility that Aurora B could be degraded dependent on p97.

2.5 Chromosomal passenger complex fate in *Xenopus laevis* egg interphase cytosol

In somatic cells, the chromosomal passenger complex is degraded quantitatively at the end of mitosis (Honda, Korner et al. 2003). To investigate whether of Aurora B and survivin are also quantitatively degraded in our system, we analysed the stability of the chromosomal passenger complex proteins in the embryonic system of *Xenopus laevis* eggs.

Xenopus laevis egg interphase cytosol was incubated with an energy mix at 19°C. Samples were taken after 0, 30, 60, 90, 120 min. The level of Aurora B and survivin in the cytosol was measured. In contrast to somatic cells, the chromosomal passenger complex is not quantitatively degraded in the embryonic system of *Xenopus laevis* eggs (Figure 18a). It stays stable over the time normal nucleus formation reaction take place. Also in the CSF extracts, which starts with high cyclin B activity in metaphase II of meiosis, the chromosomal passenger complex is stable, if the CSF extract is released into interphase by Ca^{2+} addition.

Due to the observation that the global levels of Aurora B and survivin do not change during the reaction of nucleus formation, we hypothesized that Aurora B and survivin are deubiquitinated and rescued from ubiquitination and degradation. Alternatively, only a small fraction could still be degraded in the cytosol, since the modification detected was lysine-48 chains. Therefore, we asked what the fate is of that polyubiquitinated fraction, whether it is degraded or it is deubiquitinated. To address this question, we incubated *Xenopus laevis* egg interphase cytosol with the 26S proteasome inhibitor MG132 at 0.5 mM, with 0.5 µM ubiquitin-aldehyde or with DMSO as control for 30 min. Ubiquitin-aldehyde is an inhibitor of deubiquitinating enzymes containing a enzymatic cysteine, but it does not inhibit the proteasome at this concentration (Hershko and Rose 1987). Subsequently, the three reactions were subjected to denaturing immunoprecipitations with control or Aurora B antibodies as described above (Figure 18b).

This experiment shows, that polyubiquitinated Aurora B accumulates in absence of proteasomal degradation using MG132. Since very high levels of MG132 had to be used to achieve an inhibition of the proteasome in *Xenopus laevis* egg interphase cytosol, artefacts may occur. Hence, other means of proteasomal inhibition are required to confirm this finding. By inhibition of deubiquitinating enzymes, global levels of polyubiquitinated proteins are elevated this shows that the inhibition was efficient. The levels of polyubiquitinated Aurora B were not elevated, showing that no deubiquitination of Aurora B occurs in cytosol.

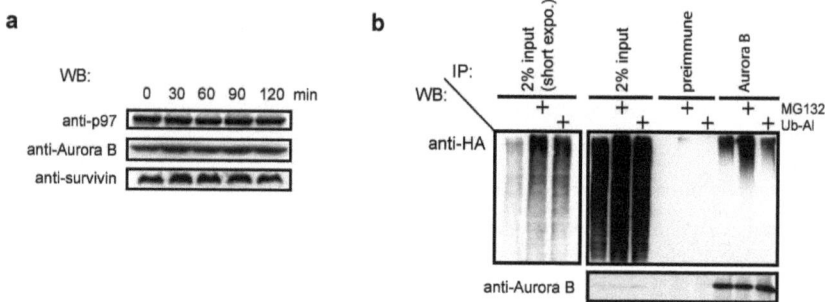

Fig. 18 The chromosomal passenger complex is not quantitatively degraded in *Xenopus laevis* egg interphase extract. a, *Xenopus laevis* egg interphase extracts were incubated with ATP-regenerating system and aliquots taken at indicated time points. Samples were analyzed by western blotting (WB) with indicated antibodies. Note that Aurora B and survivin are stable. b, *Xenopus laevis* egg interphase cytosol was incubated with 20 µM HA-tagged wt ubiquitin without or with 500 µM MG132 or 0.5 µM ubiquitin-aldehyde (Ub-Al) for 30 min. Samples were subjected to denaturing immunoprecipitation and western blotting of Aurora B and control as indicated. Inputs are also depicted at short exposure to illustrate differences. Note that ubiquitinated Aurora B accumulates only in presence of MG132.

2.6 Chromosomal passenger complex association with chromatin

Aurora B is activated, when clustered on structures like on chromatin or the kinetochores (Kelly, Sampath et al. 2007). Hence, we speculated, that the important regulation of Aurora B activity by p97 may take place when it is associated with chromatin. Therefore, we analyzed the chromatin binding of Aurora B and survivin during the process of nucleus formation in *Xenopus laevis* egg interphase extract.

We asked whether Aurora B and survivin bind to chromatin in *Xenopus laevis* egg interphase extract. To do so, demembranated sperm chromatin was incubated with *Xenopus laevis* egg interphase extract containing 8 µM p97ΔD2 wt or p97ΔD2 buffer. After 0, 5, 60 min the reaction was diluted in egg lysis buffer and recovered by centrifugation through a sucrose cushion to remove unbound material. Western blot analysis of the recovered chromatin fraction showed that Aurora B and survivin quickly bound to chromatin, full binding was reached already after 5 min (Figure 19, lane 1-5). To analyse the levels of chromatin bound Aurora B and survivin dependence of p97 activity, we performed the same reaction in presence of the dominant negative fragment p97ΔD2. Slightly higher levels of Aurora B and survivin detected on chromatin in absence of p97 activity (Figure 19, lane 6-9).

Usually about 2% of the chromosomal passenger complex proteins Aurora B and survivin bound in a standard reaction with 500 sperm heads/μl extract. Interestingly, p97 binding to chromatin was detected. The slightly elevated levels of Aurora B and survivin on chromatin in absence of p97 activity and the binding of p97 to chromatin pointed towards an involvement of p97 in Aurora B and survivin chromatin association.

Kristijan Ramadan analysed the activity of Aurora B in cytosol and on chromatin dependent on p97 activity. In agreement with these results, he showed that Aurora B has a higher activity on chromatin in absence of p97 activity, whereas in cytosol, no regulation could be detected (Ramadan, Bruderer et al. 2007).

This may be explained by the model that p97 mobilizes Aurora B from chromatin. To test if p97 mobilizes Aurora B we set up the next experiments.

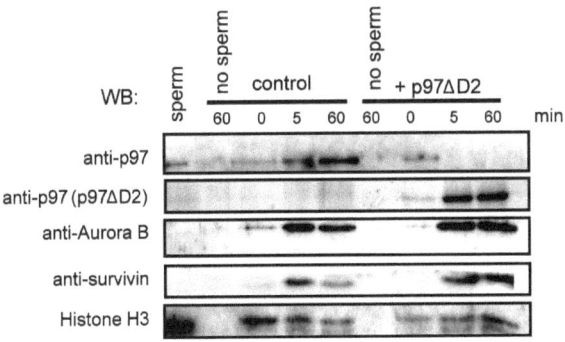

Fig. 19 Association of Aurora B and survivin with chromatin during NE formation in dependence of p97 function. *Xenopus laevis* egg interphase extracts were incubated with ATP-regenerating system and 8 μM p97ΔD2 or p97ΔD2 buffer. After the indicated time points, reactions were diluted, chromatin recovered by centrifugation through a sucrose cushion and analyzed by western blotting (WB) with indicated antibodies. Histone H3 Ponceau S stain serves as chromatin loading control.

2.7 Mobilization of Aurora B from chromatin

It has been shown that p97 mobilizes its substrate proteins from larger structures, like the unfolded polypeptide chains in ERAD or transcription factors from its unprocessed partner. We hypothesised that p97 could be responsible of reducing the levels of Aurora B on chromatin by extracting it. To measure the dissociation of Aurora B from chromatin isolated from re-association, of Aurora B, we set up a novel mobilization assay. Additionally, we want to address in the following sections the individual aspects of the process to be able to compare it to the established mechanism of p97 in ER associated processes.

2.7.1 Establishment of the Aurora B Mobilization assay

To monitor the dissociation of Aurora B from chromatin in an isolated manner, we first loaded demembranated sperm chromatin with Aurora B by incubating it with complete *Xenopus laevis* egg interphase cytosol at 19°C for 10 min. Then we re-isolated the chromatin through centrifugation through a sucrose cushion and incubated it with *Xenopus laevis* egg interphase cytosol depleted of Aurora B to prevent rebinding of soluble Aurora B (Figure 20a). Under these conditions the levels of Aurora B dropped to about 5-10 % after 60 min, compared to a second control incubation with complete cytosol, where the level of Aurora B stayed close to constant (Figure 20b). The Ponceau S stain of Histone H3 on the membrane was used as control for the amount of chromatin isolated. This shows that Aurora B association with chromatin is dynamic.

Next, we tested whether the dissociation of Aurora B from chromatin was energy-dependent. Therefore, the assay was performed as described before, but the second incubation was performed in *Xenopus laevis* egg interphase cytosol depleted of Aurora B in presence of ATP or absence of ATP and addition of 0.1 units/µl apyrase. In absence of ATP, the dissociation of Aurora B from chromatin was slowed down (Figure 20c). The MCM3 western blot signal was used as control of the amount of chromatin isolated. The levels of MCM3 are also affected by the lack of ATP since it is an ATPase and its chromatin association is ATP dependent (Maiorano, Lemaitre et al. 2000). However, the Ponceau S stain of histone H3 proved that equal amounts were loaded.

These results show that the association of Aurora B to chromatin is dynamic. Interestingly, the mobilization of Aurora B from chromatin requires energy. In the investigated processes p97 mobilizes its substrates in an energy dependent manner. Therefore, we asked whether the energy-dependent mobilization requires p97.

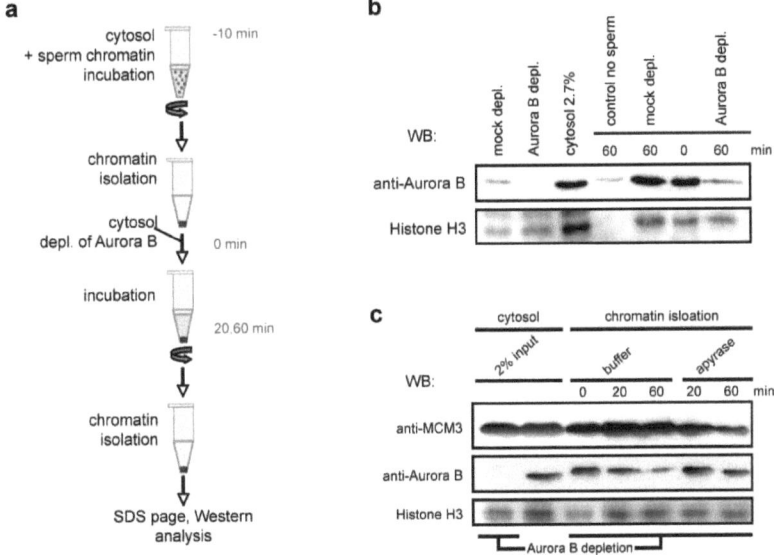

Fig. 20 Mobilization assay: ATP-dependent dissociation of Aurora B from chromatin. a, scheme of the Aurora B mobilization assay: Sperm chromatin was incubated with complete cytosol to load it with Aurora B and isolated through a sucrose cushion (0 min). It was then incubated with Aurora B depleted cytosol, and re-isolated at indicated times for analysis by western blotting with antibodies or Ponceau stain (H3). b, mobilization assay performed as in (a), the second incubation step was done either with mock- or Aurora B depleted extract. The chromatin was isolated after the indicated time points and analysed by western blotting (WB). c, mobilization assay as in (a) but in the second incubation step performed as control or with supplement of 0.1 units/µl apyrase. The chromatin was isolated after the indicated time points and analysed by western blotting. MCM3 and Histone H3 were probed to control chromatin recovery.

2.7.2 Mobilization of Aurora B from chromatin is dependent on p97

Since we detected an energy dependent mobilization, we asked whether the energy dependent mobilization of Aurora B from chromatin required p97. We used the mobilization assay to address this question. The p97 function was inhibited in three independent ways, either by depletion of p97, by using the dominant negative fragment p97ΔD2 wt and p97ΔD2 K251A as control or by using the inhibitory monoclonal antibody against Ufd1.

Inactivation of p97 by depletion

The first approach was using double depleted (Aurora B and p97) cytosol for the second incubation in the mobilization assay (Figure 20a). We first loaded demembranated sperm chromatin with Aurora B by incubating it with complete *Xenopus laevis* egg interphase cytosol. Then we re-isolated the chromatin through centrifugation through a sucrose cushion and incubated it with *Xenopus laevis* egg interphase depleted of Aurora B as before and additionally either mock or p97 complexes depleted with GST or GST-UT6, respectively. The chromatin was isolated by centrifugation through a sucrose cushion after 0, 20 and 60 minutes incubation and analysed by western blotting (Figure 21a). In the input lanes (1-3), the degree of depletion is analysed. In the last lane, the cytosol contamination for the chromatin isolation step was monitored. The Ponceau S stain of Histone H3 and the MCM3 western blot signal were used as control of the amount of chromatin isolated. Strikingly, in absence of p97, the mobilization of Aurora B from chromatin was strongly reduced. Detection of the western blot signal was done using chemiluminescence and quantification was performed directly with an AlphaInnotech Fluorchem 8900 imager for three independent experiments (Figure 21b).
Depletion of p97 resulted in a strong retardation of mobilization of Aurora B compared to mock depletion.

Dominant negative approach to inactivate p97

In the second approach, we used the method of the dominant-negative p97 fragment p97ΔD2. Taking advantage of the mobilization assay (Figure 20a), we loaded demembranated sperm chromatin with Aurora B by incubating it with complete *Xenopus laevis* egg interphase cytosol and re-isolated the chromatin as before. Subsequently, we incubated it with *Xenopus laevis* egg interphase cytosol depleted of Aurora B containing 8 μM p97ΔD2 wt, 8 μM p97ΔD2 K251A or p97ΔD2 buffer. The chromatin was isolated by centrifugation through a sucrose cushion after 0, 10 and 30 minutes and analysed by western blotting (Figure 21c). The Ponceau S stain of Histone H3 and the MCM3 western blot signal were used as control of the amount of chromatin isolated. The dissociation of Aurora B from chromatin was delayed in presence of p97ΔD2 wt compared to the controls buffer and p97ΔD2 K251A. Detection of the western blot signal was done using chemiluminescence and quantification was performed directly with an AlphaInnotech Fluorchem 8900 imager for three independent experiments (Figure 21d).

The inactivation of p97 by the dominant negative p97ΔD2, but not the negative fragment p97ΔD2 K251A nor the buffer control showed retarded mobilization.

Inactivation of p97 by addition of an inhibitory antibody against Ufd1
The third approach was interfering with p97$^{Ufd1-Npl4}$ activity by usage of the inhibitory monoclonal anti-Ufd1 antibody. We again used the mobilization assay (Figure 20a). We first incubated demembranated sperm chromatin with complete *Xenopus laevis* egg interphase cytosol and re-isolated the chromatin as before. Then, we incubated it with *Xenopus laevis* egg interphase cytosol depleted of Aurora B containing 0.62 µg/µl anti-Ufd1 monoclonal antibody 5E2 in egg lysis buffer, egg lysis buffer as control or in presence of 0.1 units/µl apyrase. The chromatin was re-isolated by centrifugation through a sucrose cushion after 0, 10 and 30 minutes and analysed by western blotting (Figure 21e). The Ponceau S stain of Histone H3 and the MCM3 western blot signal were used as control of the amount of chromatin isolated. The mobilization of Aurora B from chromatin was strongly delayed in the presence of the anti-Ufd1 monoclonal antibody 5E2 compared to the presence of its buffer. The presence of apyrase even stronger inhibited mobilization of Aurora B. The Quantification was preformed directly with an AlphaInnotech Fluorchem 8900 imager (Figure 21f).
Addition of the inhibitory anti-Ufd1 antibody blocked the mobilization of Aurora B.

All three different approaches to inactivate p97 led to a retarded mobilization of Aurora B from chromatin. This demonstrates that p97$^{Ufd1-Npl4}$ mobilizes Aurora B from chromatin in an energy dependent manner, as it has been reported for p97$^{Ufd1-Npl4}$ function in ERAD and transcription activation. In the process of ERAD, the inhibition of p97 results in accumulation of polyubiquitinated substrates on the ER membrane. Therefore, we asked if polyubiquitinated Aurora B accumulates on chromatin in absence of p97 function.

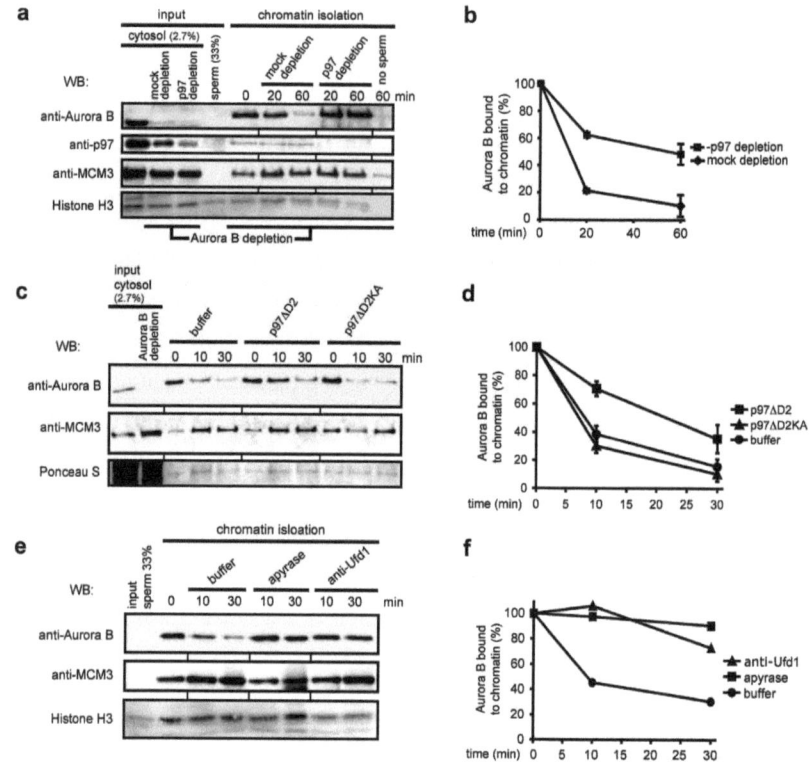

Fig. 21 p97 extracts Aurora B from chromatin. a, sperm chromatin was incubated with complete *Xenopus laevis* egg interphase cytosol to load it with Aurora B and next isolated through a sucrose cushion (0 min). Then, it was incubated with Aurora B depleted cytosol that was either mock- or p97-depleted, and re-isolated at indicated times for analysis by western blotting (WB) with antibodies or Ponceau stain (H3). **b**, quantification of western blotting signal (n=3, bars show s.e.m.). **c**, sperm chromatin was incubated with complete *Xenopus laevis* egg interphase cytosol to load it with Aurora B and isolated through a sucrose cushion (0 min). The sperm chromatin was then incubated with Aurora B depleted cytosol containing 8 μM p97ΔD2 wt, 8 μM p97ΔD2 K251A or p97ΔD2 buffer, and then re-isolated at indicated times for analysis by western blotting with antibodies or Ponceau stain (H3). **d**, quantification of western blot signal (n=3, bars show s.e.m.). **e**, the experiment was carried out as in (c) in the absence or presence of anti-Ufd1 monoclonal antibody 5E2 (0.62 μg/μl), and then re-isolated at indicated times for analysis by western blotting with antibodies or Ponceau stain (H3). **f**, quantification of western blotting signal.
MCM3 and Histone H3 were probed to control chromatin recovery.

2.7.3 Ubiquitinated Aurora B accumulates on chromatin in absence of p97 function

It has been reported for the p97 dependent mobilization in ERAD, that in absence of p97 function ubiquitinated substrates of p97 accumulate on the ER membrane (Ye, Meyer et al. 2003). We analysed whether this is also true in the Aurora B mobilization from chromatin. To do so, we incubated demembranated sperm chromatin in *Xenopus laevis* egg interphase cytosol containing 20 µM HA-tagged ubiquitin and 8 µM p97ΔD2 or an equivalent volume of storage buffer at 19°C for 30 min. Then, the reactions were stopped by cooling on ice and addition of 5 mM N-ethylmaleimide and the chromatin was isolated. 2.5% of the chromatin was directly analysed by western blotting. From 97.5% of the reactions the proteins were eluted from chromatin with 600 mM NaCl. The eluates were subjected to immunoprecipitation with control preimmune or Aurora B antibodies. The anti MCM3 signal is used as loading control for chromatin (Figure 22).

Analysis by western blotting revealed that ubiquitinated Aurora B can be detected on chromatin. Furthermore, inactivation of p97, by addition of the dominant negative p97 fragment p97ΔD2, resulted in accumulation ubiquitinated Aurora B on chromatin. This is in agreement with the results obtained in the ERAD process. Ubiquitinated ERAD substrates accumulate at the ER membrane in absence of p97 activity.

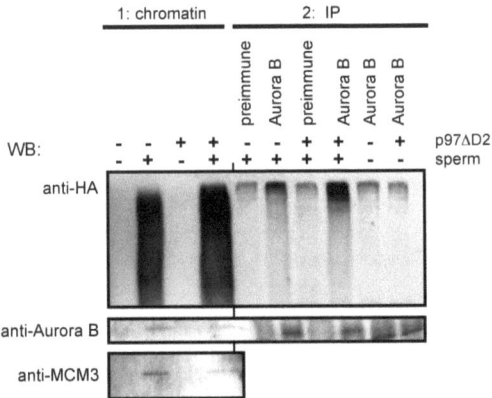

Fig. 22 Ubiquitinated Aurora B accumulates on chromatin in absence of p97 function. Chromatin was incubated in *Xenopus laevis* egg interphase cytosol containing 20 µM HA-tagged ubiquitin and 8 µM p97ΔD2 or p97ΔD2 buffer, and then re-isolated; 2.5% of the reactions was directly analysed by western blotting (WB) (1: chromatin). From the residual chromatin, proteins were salt-eluted, subjected to Aurora B or control immunoprecipitations, and the isolates analysed as indicated (2: IP).

2.8 Mobilization of Aurora B from chromatin influenced by proteasome activity

During ERAD, proteasomal activity is required to mobilize substrates from the ER-membrane. In absence of proteasome activity degradation substrates accumulate on the ER membrane. Thus, we analysed whether the proteasome influences the mobilization of Aurora B from chromatin. To analyse this, we used the mobilization assay (Figure 20a).

We first incubated demembranated sperm chromatin with complete *Xenopus laevis* egg interphase cytosol to associate Aurora B with chromatin. Then, we re-isolated the chromatin through centrifugation through a sucrose cushion and incubated it with *Xenopus laevis* egg interphase cytosol depleted of Aurora B containing 500 µM MG132 in DMSO or DMSO only. The chromatin was re-isolated by centrifugation through a sucrose cushion after 0, 20 and 60 minutes and analysed by western blotting. The Ponceau S stain of Histone H3 and the MCM3 western blot signal were used as control of the amount of chromatin isolated (Figure 23).

The mobilization of Aurora B from chromatin was delayed in the presence of MG132 compared to the control. This shows that proteasomal inactivation results in accumulation of unmodified Aurora B on chromatin. This finding enforces the hypothesis, that p97 mobilizes ubiquitinated Aurora B from chromatin for degradation in an analogues way as it mobilizes substrates in ERAD.

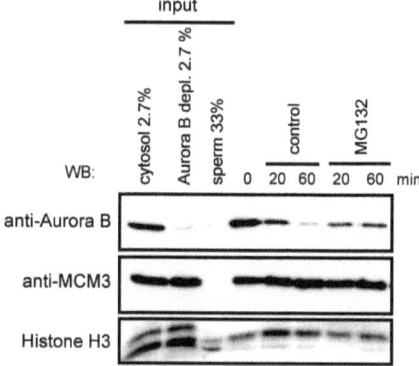

Fig. 23 Mobilization of Aurora B from chromatin is delayed in presence of MG132. Sperm chromatin was incubated with complete *Xenopus laevis* egg interphase cytosol to load it with Aurora B and isolated through a sucrose cushion (0 min). The sperm chromatin was then incubated with Aurora B depleted cytosol containing 500 µM MG132 or DMSO, and then re-isolated at indicated times for analysis by western blotting with antibodies or Ponceau stain (H3). MCM3 and Histone H3 were probed to control chromatin recovery.

2.9 Interaction of the ligase Cullin3 with Aurora B and p97

We have detected lysine-48 chain linked polyubiquitinated Aurora B in *Xenopus laevis* egg interphase cytosol and bound to chromatin. Therefore, an obvious question was, the identity of the ligase that mediates the polyubiquitination of Aurora B. Recently, it was reported that in somatic human cells the multi subunit ligase Cullin3 with the substrate adaptors KLHL9 and KLHL13 polyubiquitinates Aurora B in mitosis (Sumara, Quadroni et al. 2007). The authors further found that Aurora B binds to the substrate recognition domain of KLHL9 and KLHL13. Additionally, they could show that Cullin3 with KLHL9 and KLHL13 polyubiquitinates Aurora B *in vivo* and *in vitro*. Cullin3 is a multi protein ligase with multiple substrate adaptors and therefore is involved in multiple processes in the cell. Neddylation and deneddylation are required for its function (Bosu and Kipreos 2008). We asked whether Cullin3 also in *Xenopus laevis* egg extracts was the ligase responsible for the polyubiquitination of Aurora B. In *Xenopus laevis* databases we could identify only one gene closely homologue to the KLHL9 and KLHL13 adaptors. Since they are highly conserved in sequence, it is possible that in *Xenopus laevis*, the single form substitutes the two in the human (Accession No.: MGC80367). We term it here KLHL913. The substrate adaptors contain an N-terminal BTB domain, which binds to Cullin3. The Kelch repeats at the C-terminus mediate substrate binding (Figure 24a).

First, we wanted to reproduce the interaction of Aurora B with Cullin3. Therefore, we performed native immunoprecipitations from *Xenopus laevis* egg interphase cytosol for Cullin3 and control antibodies. Western blot analysis showed an interaction of Cullin3 with Aurora B, with KLHL913 and with p97 (Figure 24b).

This showed that the interaction between Cullin3 and Aurora B is conserved in the *Xenopus laevis* system. Therefore, we tested if Cullin3 ubiquitinates Aurora B in *Xenopus laevis* egg interphase cytosol.

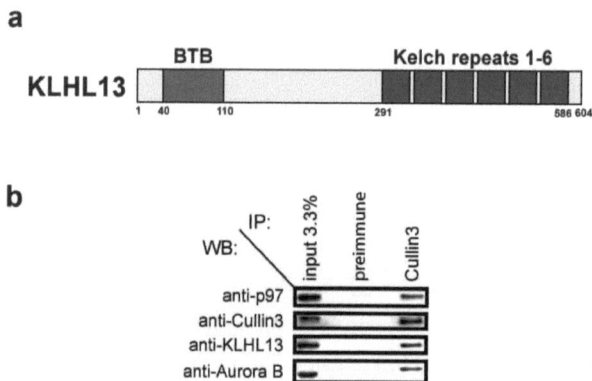

Fig. 24 KLHL13 domain structure and interaction of Cullin3 with Aurora B, p97 and KLHL913. a, Domain layout of human KLHL13 depicted with the BTB domain that interacts with Cullin3 and the Kelch repeats which interact with the substrates. b, Cullin3 was immunoprecipitated (IP) from *Xenopus laevis* egg interphase cytosol under native conditions using the purified antibodies and preimmune antibodies. The isolates were analysed by western blotting (WB).

2.10 Analysis of polyubiquitination of Aurora B by Cullin3

2.10.1 Analysis of ubiquitination of Aurora B by Cullin3 using immunodepletion

Since we could reproduce the interaction of Cullin3 with Aurora B in *Xenopus laevis* egg interphase cytosol, we asked, if Cullin3 generates the polyubiquitin chain modification of Aurora B. We addressed this question for the cytosolic pool of Aurora B, since there we see polyubiquitination and a dependence of p97 for the polyubiquitination. We first tried to deplete the ligase Cullin3 from cytosol and then test if polyubiquitination of Aurora B still occurs.

We depleted Cullin3 from *Xenopus laevis* egg interphase cytosol using and anti-Cullin3 antibody. In western blotting, this antibody recognizes two distinct bands at the size of Cullin3, most likely the unmodified and the neddylated form of Cullin3. Interestingly, this antibody was only able to deplete the lower band (Figure 25a). We used the mock and Cullin3-depleted cytosol and incubated it with 20 µM HA-tagged ubiquitin for 30 min. Subsequently, the reactions were subjected to denaturing immunoprecipitations with control or Aurora B antibodies as described above (Figure 25b).

No difference was observed in polyubiquitination between the mock and the Cullin3-depleted reaction. Since the depletion was successful only for one form of Cullin3, it remains the possibility that the remaining Cullin3 is sufficient for polyubiquitination of Aurora B. Therefore, we developed other tools to inhibit Cullin3.

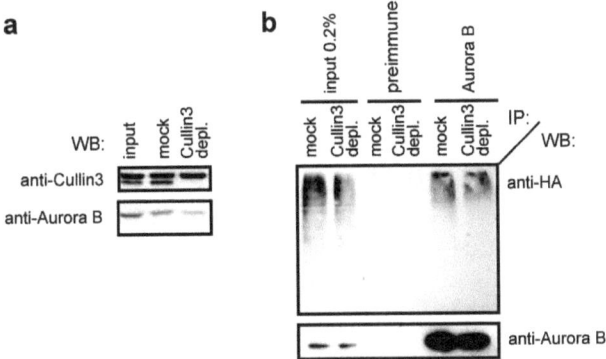

Fig. 25 Analysis of polyubiquitination of Aurora B by Cullin3.a, *Xenopus laevis* egg interphase cytosol was mock- or Cullin3-depleted. Analysis was performed by western blotting (WB). b, the cytosols from (a) were supplemented with 20 μM HA-tagged wt ubiquitin and incubated for 30 min, denatured and subjected to immunoprecipitation (IP) of Aurora B, INCENP or with preimmune antibodies. Polyubiquitination was detected by western blotting with HA-specific antibodies.

2.10.2 Analysis of polyubiquitination of Aurora B by Cullin3 using depletion with a fragment of KLHL13

To have other tools to inhibit Cullin3 we generated a fragment of its substrate adaptor KLHL13. This fragment of KLHL13 comprises the binding site for Cullin3 but not for Aurora B and thereby could act dominant negatively or it could be used to specifically deplete Cullin3 from cytosol. Additionally, we generated a mutated version of the KLHL13 fragment, in which the putative binding sites for Cullin3 are mutated. To identify these residues, we compared the BTB of KLHL13 with the BTB of SKP1, which binds to Cullin1 and a mutated version of the BTB interfering with the Cullin1 binding has been successfully created (Xu, Wei et al. 2003). To be able to compare, the structure of the BTB domain of KLHL13 was modelled with a template based approach using the SWISS-MODEL server (http://swissmodel.expasy.org/SWISS-MODEL.html). The modelled BTB domain of KLHL13 was compared with the structure of the BTB domain of SKP1 (Schulman, Carrano et al. 2000) to identify the corresponding residues. These residues were identified as being

F173A and L174A, the fragment is termed KLHL13 FL173AA (Figure 26a). It serves as a control for the KLHL13 wt fragment, as it may be only negative.

We cloned and expressed His-tagged KLHL13 wt and KLHL13 FL173AA fragments. These fragments were purified via Ni^{2+} affinity and analysed by SDS-PAGE. The purification resulted in strongly enriched protein fractions (Figure 26b).

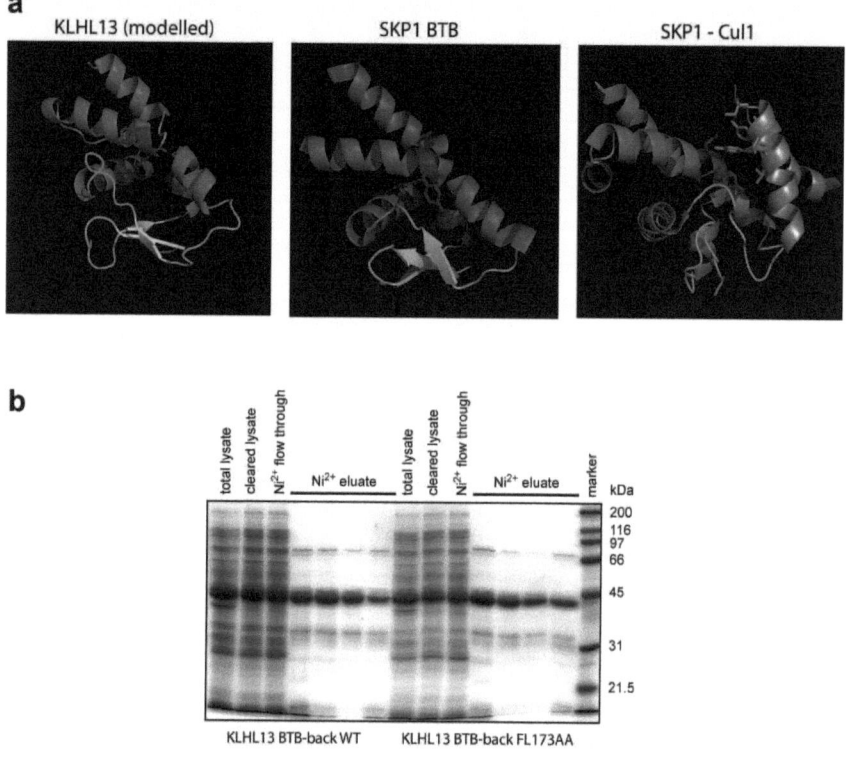

Fig.26 KLHL13 wt and KLHL13 FL173AA fragments. a, left is the modelled structure of KLHL13 BTB domain with the residues FL173 coloured in blue. In the middle is the structure of the SKP1 BTB with the residues NY108 mutated to interfere with Cullin1 interaction depicted in blue. The right model shows the interface of interaction between Cullin1 and Skp1. **b,** His-tagged KLHL13 wt and KLHL13 FL173AA fragments (containing the BTB and Back domains) were expressed in *E. coli* and purified via Ni^{2+} affinity. Coomassie stained SDS-PAGE gels of Ni^{2+} affinity purification fractions shown.

We used the purified His-tagged KLHL13 wt and KLHL13 FL173AA to deplete Cullin3 from *Xenopus laevis* egg interphase cytosol. The KLHL13 wt fragment was able to deplete Cullin3, but only the upper supposedly neddylated form of Cullin3. The KLHL13 FL173AA fragment did not deplete any band of Cullin3 (Figure 27a). Interestingly, the wt fragment also depleted the endogenous KLHL913 adaptor of Cullin3 from the cytosol (Figure 27c, lower panel). The same was also the case with the KLHL13 FL173AA fragment. The reason for this could be the dimerisation of the endogenous KLHL913 protein with the exogenous fragment. This means that the fragment is most likely not just a negative, but interferes with the endogenous KLHL913 and therefore with the function of Cullin3.

The mock and the Cullin3 depleted cytosols were incubated with 20 µM HA-tagged ubiquitin for 30 min. Subsequently, the reactions were subjected to denaturing immunoprecipitations with control or Aurora B antibodies as described above (Figure 27b). No difference in polyubiquitination could be observed between the mock and the depleted with KLHL13 wt or KLHL13 FL173AA reactions.

Since an increased ubiquitination of Aurora B was observed in absence of p97 function, we hypothesised, that the fraction of Aurora B ubiquitinated by Cullin3 could only be visible in absence of p97 activity. Therefore, we performed the experiment as in Figure 27b, but in presence of 8 µM p97ΔD2 (Figure 27d). Strikingly, the ubiquitination of Aurora B was reduced in the reactions depleted with KLHL13 wt or KLHL13 FL173AA. This suggests that Cullin3 is partly responsible for Aurora B ubiquitination in *Xenopus laevis* egg interphase cytosol, but additional controls are required. The specificity of the depletion needs to be shown by reconstitution of the depleted reaction with recombinant Cullin3.

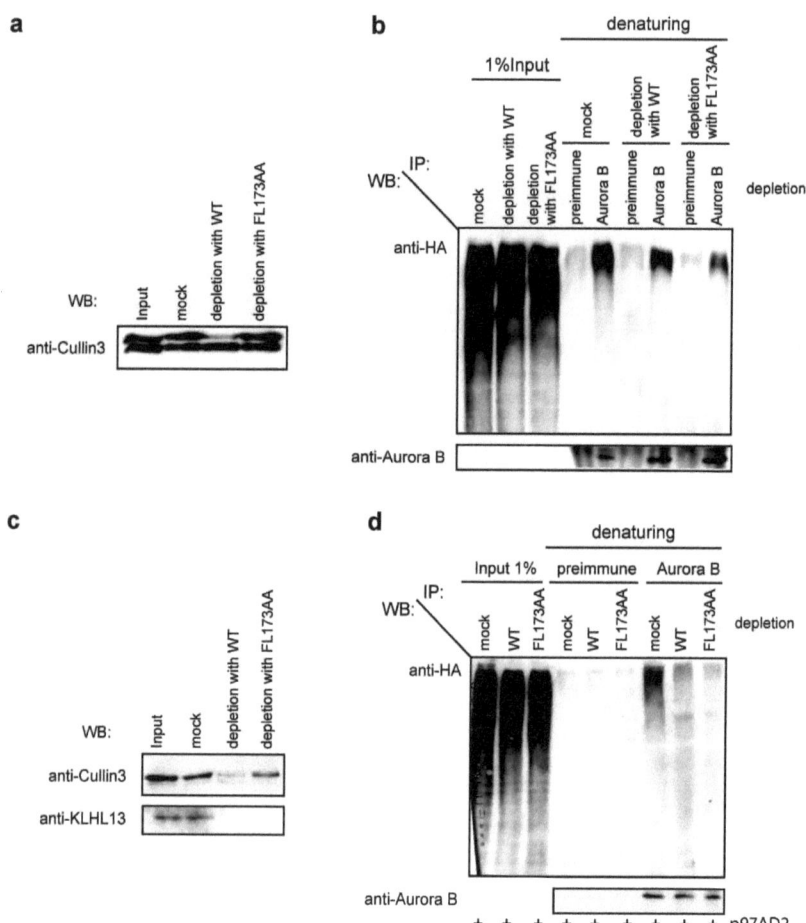

Fig. 27 Analysis of the ubiquitination state of Aurora B by Cullin3 depletion via its adaptor.
a, *Xenopus laevis* egg interphase cytosol was mock-depleted, KLHL13 wt fragment depleted or KLHL13 FL173AA depleted. Analysis of depletions was performed by western blotting (WB). b, the cytosols from (a) were supplemented with 20 μM HA-tagged wt ubiquitin and incubated for 30 min, denatured and subjected to immunoprecipitation of Aurora B, with preimmune antibodies. Polyubiquitination was detected by western blotting with HA-specific antibodies. c, depletions done as in (a). d, reaction done as in (b) but in presence of 8 μM p97ΔD2.
The efficiency of Aurora B isolation in (a) and (b) was controlled with specific antibodies.

2.10.3 Analysis of polyubiquitination of Aurora B by Cullin3 using a dominant negative KLHL13 fragment

Alternatively, to the depletion of Cullin3 we performed the inhibition of the Cullin3 by addition of the KLHL13 wt fragment as a dominant-negative, supposedly by titrating endogenous Cullin3 away from its substrates. As control of the specificity, KLHL13 FL173AA was used. We incubated *Xenopus laevis* egg interphase cytosol containing 20 µM HA-tagged with 7.8 µM KLHL13 wt or 7.8 µM KLHL13 FL173AA or buffer for 30 min. Subsequently, the reactions were subjected to denaturing immunoprecipitations with control or Aurora B antibodies as described above (Figure 28a). No difference could be observed under these conditions, the ubiquitination status of Aurora B remained unchanged upon addition of the KLHL13 fragments.

As before, we added the dominant negative fragments in absence of p97 function. To analyse this, we performed the experiment as in Figure 28a, but in presence of 8 µM p97ΔD2 (Figure 28b). Importantly, the ubiquitination of Aurora B was reduced in the reactions containing KLHL13 wt. A reduction of the levels of polyubiquitinated Aurora B was also observed in the KLHL13 FL173AA reaction. An explanation for this may be that KLHL13 FL173AA could also act dominant negative by interacting with endogenous KLHL913 and thereby interfering with the ligase substrate complex (see figure 27c). These findings suggest that Cullin3 via KLHL913 is partly responsible for Aurora B ubiquitination in *Xenopus laevis* egg interphase cytosol, but more controls are required, since we have no control fragment that is proven only negative.

The results from the depletion and the dominant negative experiments show that the Aurora B is ubiquitinated by Cullin3 based ligase complexes, but only a part of the total ubiquitination of Aurora B is dependent on Cullin3 namely the part which is regulated by p97. It remains obscure which ligase is responsible for the Cullin3 independent ubiquitination. Further experiments are required for understanding of this process.

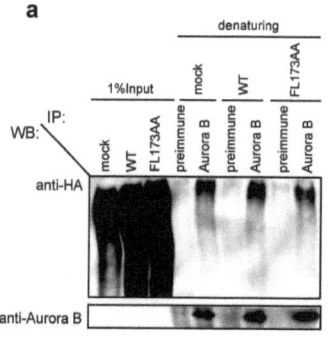 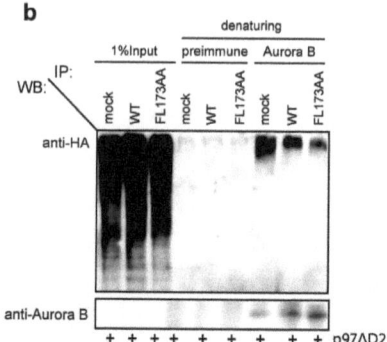

Fig. 28 Analysis of the ubiquitination of Aurora B by Cullin3 using a dominant negative KLHL13 fragment. a, *Xenopus laevis* egg interphase cytosol was incubated with 20 μM HA-tagged wt ubiquitin and 7.8 μM KLHL13 wt fragment or KLHL13 FL173AA fragment or KLHL13 buffer for 30 min. Subsequently, the reactions were subjected to denaturing immunoprecipitation of Aurora B or with preimmune antibodies. **b**, *Xenopus laevis* egg interphase cytosol was incubated with 20 μM HA-tagged wt ubiquitin, 8 μM p97ΔD2 and 5.9 μM KLHL13 wt fragment or KLHL13 FL173AA fragment or KLHL13 buffer for 30 min. Next, the reactions were subjected to denaturing immunoprecipitation of Aurora B or with preimmune antibodies.
The efficiency of Aurora B isolation in (a) and (b) was controlled with specific antibodies.

2.11 Regulation of the ATPase activity of p97 by p47 and Ufd1-Npl4

The mobilization of ubiquitinated Aurora B from chromatin by p97 required the substrate adaptor Ufd1-Npl4 and ATP hydrolysis of p97. It was reported that p47 if in complex with p97, reduced the ATPase activity of p97 to 15% (Meyer, Kondo et al. 1998). Therefore, we investigated the effect of Ufd1-Npl4 binding to p97 on the ATPase activity. Ufd1-Npl4 is a heterodimeric protein, we found, that it exhibits to p97 the same bipartite binding mechanism to p97 like p47. Npl4 contains an ubiquitin fold domain (UBD) that binds like the UBX domain of p47 to the N-domain of p97. Furthermore, there is a second binding site in Ufd1 called BS1 that binds also to the N-domain of p97. p47 does have the same BS1 (Figure 29) (Bruderer, Brasseur et al. 2004). We hypothesised that the adaptor Ufd1-Npl4 could also downregulate the ATPase activity of p97, if the regulation of p97 by p47 is caused by the bipartite interaction mechanism.

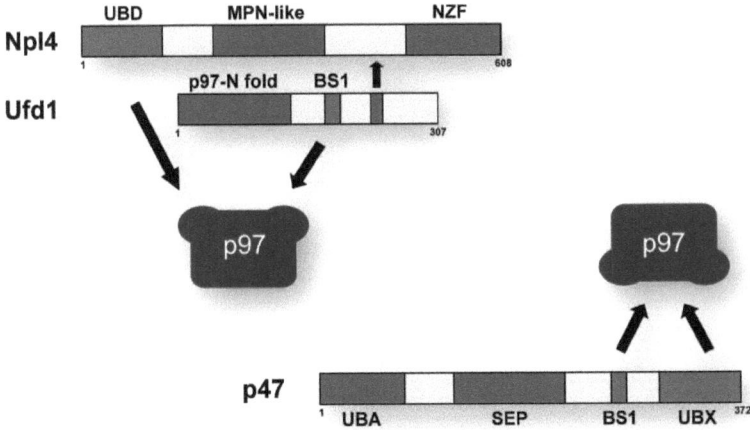

Fig. 29 Bipartite binding mechanism of p47 and Ufd1-Npl4 for p97.

To experimentally address this question, we set up the malachite green assay (modified from (Lanzetta, Alvarez et al. 1979)). This assay measures the formation of orthophosphate, Pi, generated by the ATP hydrolysis to ADP + Pi by the ATPase. Pi forms a complex with molybdate to phosphomolybdenum, and phosphomolybdenum complexes with malachite green, generating an absorbance shift of the solution. This shift is visible in the reaction tube. Without the presence Pi, the reaction has a yellow colour and if Pi is generated turns into a green blue. The progress of the reaction is measured by absorbance at 650 nm (Figure 30).

The ATPase was incubated with ATP in a suitable buffer. After different time points, aliquots of the reaction were taken and mixed with the malachite green reagents. This stopped all ATPase activity and the generated Pi complexed with the malachite green reagents. After 10 min, the absorbance was measured. The magnitude of absorbance in the malachite green reagents upon addition of Pi was calibrated using a Pi standard solution.

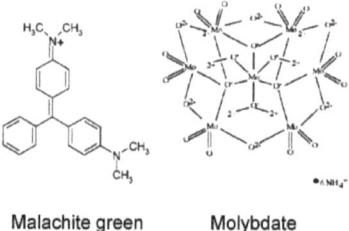

Fig. 30 Malachite green assay to measure the ATPase activity of p97. a, the Malachite green assay measures the formation of orthophosphate, Pi, upon hydrolysis, based on complex formation between phosphomolybdenum and malachite green. The reaction progress was measured by absorbance at 650 nm.

The bipartite interaction domains do not mediate regulation of p97 directly

After having established the assay to measure p97 ATPase activity, we addressed the regulation of p97 by its adaptors. p97 at 0.02 µg/µl was incubated with equal molar amounts of p47 or Ufd1-Npl4 on ice for 30 min. The ATP hydrolysis activity of the complexes was then measured at 2 mM ATP at 37 °C. p97 alone displayed at ATPase activity of 19 µmol P_i/mg p97/h. The ATPase activity of p97 was reported to generate about 15 µmol P_i/mg p97/h at 2 mM ATP concentration at 37°C at pH: 7.4 (Meyer, Kondo et al. 1998). We could reproduce the reduction of the ATPase activity of p97 by p47 that was detected earlier (Meyer, Kondo et al. 1998). Surprisingly, the presence of equal molar concentrations of Ufd1-Npl4 did not influence the ATPase activity of p97. We tested higher molar amounts of Ufd1-Npl4 with p97 and could detect a reduction of the ATPase activity of p97 (data not shown).

We hypothesized, that the bipartite interaction domains per se does not mediate directly the inhibition detected by the p97-p47 complex, since the p97-Ufd1-Npl4 complex has the same bipartite interaction motif. So we concluded that additional domains in p47 could be responsible for the inhibition of the ATPase activity of p97.

To address the hypothesis that additional domains in p47 are responsible for the reduced ATPase activity or the p97-p47 complex, we created a His-tagged fragment of p47 containing only BS1 and the UBX domain. The fragment ranged from amino acids 244-370 and was termed p47 (244–370). Note in this fragment are comprised the UBA and the SEP domain.

Recombinant His-p47 (244–370) was expressed in *E.coli*. The first purification step was Ni^{2+} affinity chromatography. Subsequently, the eluate peak was purified by gelfiltration chromatography using a Superdex 200 column. The resulting protein was highly pure.

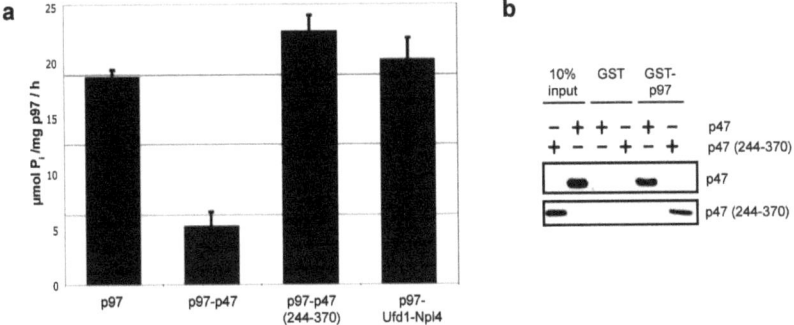

Fig. 31 Bipartite interaction domains do not mediate regulation of p97 directly. a, p97 complexes were formed by preincubating p97 with equal molar amounts of p47, the p47 (244–370) fragment containing both BS1 and UBX, or Ufd1-Npl4 on ice for 30 min. The ATP hydrolysis activity of the complexes was then measured at 37 °C. Shown is the mean of three experiments. **b**, equal binding affinities of p47 and p47 (244–370) to p97 were confirmed by co-precipitation experiments. GST or GST-p97 was incubated with either of the two His-tagged p47 variants, and bound protein was analyzed by immunoblotting using an anti-His antibody.

Next we tested p47 (244–370) fragment in the malachite green assay. No ATPase activity was detected, meaning absence of contaminating ATPase activity. Now, we could test, whether p47 (244–370) regulated the ATPase activity of p97 as p47 full-length did. p97 complexes were formed by preincubating p97 with equal molar amounts of p47, the p47 (244–370) fragment, or Ufd1-Npl4 on ice for 30 min. The ATP hydrolysis activity of the complexes was then measured at 37 °C. Shown is the mean of three experiments (Figure 31a). The p97-p47 (244–370) complex did not display reduced ATPase activity. This shows that the bipartite interaction mechanism is not responsible directly for the down regulation of the ATPase activity of p97. One possibility, why p47 (244–370) did not down regulate the ATPase activity of p97 is, that it did not bind to p97. To rule out the possibility that the p47 (244–370) fragment has a lower affinity to p97 than full-length p47, we performed pull down experiments. GST and GST-p97 were incubated with full-length p47 or with p47 (244–370). Then, GSH Sepharose beads were used isolate GST or GST-p97 (Figure 31b). Full length p47 and p47

(244–370) showed equal affinities to p97. This confirms that p47 (244–370) bound to p97 but did still not suppress its ATPase activity and confirms that the binding module of p47 alone does not regulate the ATPase activity of p97 upon binding. Additional domains of p47 or its trimerisation may contribute to the suppression of the ATPase activity of p97. Ufd1-Npl4 oligomerization has not been detected.

3. Discussion

The results presented in this thesis provide evidences for a novel crucial pathway that regulates reformation of the nucleus after mitosis in embryonic cells. The evidence suggests that, during mitosis, Aurora B inhibits formation of the nucleus by preventing chromatin decondensation and nuclear envelope formation. During the exit from mitosis, the ubiquitin-selective chaperone complex p97$^{Ufd1-Npl4}$ inactivates Aurora B on chromatin, thus allowing Aurora B substrates to be dephosphorylated and the nucleus to assemble. Inactivation of Aurora B is achieved first by polyubiquitination of Aurora B on chromatin and then by p97$^{Ufd1-Npl4}$ dependent mobilization in an energy dependent manner. Our results identify Aurora B as the first ubiquitinated substrate of p97 in mitosis that is relevant in biogenesis of an organelle.

3.1 Functional Role of p97 in nucleus formation

Several functions of p97 in mitosis have been discovered. Common to these finding is that p97 is a positive regulator in all of these processes. For example p97 stabilizes Cut1/separase during anaphase in a proteasome independent manner (Ikai and Yanagida 2006). During Golgi reformation after mitosis, p97 is involved in reformation of functional Golgi cisternae in a similar way as NSF is (Kondo, Rabouille et al. 1997), (Shorter and Warren 1999). It was reported that p97, via Ufd1, is involved in recruiting survivin to kinetochores in prometaphase and metaphase in somatic cells (Vong, Cao et al. 2005).
In all these processes, no direct evidences for substrates and mechanisms have been discovered. For example, no direct interaction of p97 with Cut1 has been detected, so it is unclear if the stabilization is direct. During Golgi reformation the mechanism remains unclear, although an involvement of p97 in SNARE disassembly was proposed. The recruitment of survivin to kinetochores was shown by RNAi of Ufd1 only. These mechanisms of p97 function are inconsistent with the findings of p97 function in ERAD. The ERAD process has been extensively studied and p97 has a negative role by extracting unfolded polypeptides out of the ER membrane for proteasomal destruction (Ye, Meyer et al. 2003). The function of p97 in mitotic processes has to be studied in more detail to understand these on the first sight contradictory roles of p97. Here we present insight into the molecular mechanism of the function of p97 in nucleus reformation at the end of mitosis.

3.2 $p97^{Ufd1-Npl4}$ does not mediate membrane fusion during NEF

We show in this work that p97 is a regulator and not a homotypic membrane fusion factor similar to NSF during nucleus formation. NSF disassembles t- and v-SNARE complexes after fusion in an energy dependent manner (Hong 2005). The observation, that $p97^{Ufd1-Npl4}$ activity is not required in nucleus formation in absence of Aurora B activity rules out the possibility that $p97^{Ufd1-Npl4}$ is involved directly in an essential SNARE-dependent membrane fusion. Our results show that the function of $p97^{Ufd1-Npl4}$ in nucleus formation is an analogous mechanism to its function in ERAD.

Earlier observations of p97 function in membrane fusion dependent processes led to the formulation of the hypothesis that p97 with p47 may be involved directly in homotypic SNARE-dependent membrane fusion. The conclusions were drawn from results of Golgi reassembly experiments. Additionally, the hypothesis was based on the similarity of the NSF-function in the secretory pathway and the p97 system. The two proteins display similar domain structures. Both AAA ATPases bind the substrates via N-terminally bound substrate adaptors. It was even reported that p47 competes syntaxin5 binding with alpha-SNAP (Rabouille, Kondo et al. 1998) and some evidence was found that VCIP135 may disassemble p97/p47/syntaxin5 complexes in vitro (Uchiyama, Jokitalo et al. 2002).

On the other hand newer findings make it unlikely that p97 is involved directly in SNARE-dependent membrane fusion. No requirement for p97 could be found in the homotypic fusion of immature secretory granules, a process which is dependent on NSF (Urbe, Page et al. 1998). Additionally, recent findings clarify that there is no specialized 'homotypic' fusion factor required for homotypic membrane fusion, but the normal NSF fusion machinery (Wickner and Haas 2000). Importantly for p97 role in Golgi reformation, it has been found that cis-SNARE complex dissociation is already completed by NSF at the end of Golgi fragmentation. Golgi reformation from mitotic fragments can be mediated by NSF variants that can bind but not hydrolyze ATP (Muller, Shorter et al. 2002). This shows that no disassembly of SNAREs is required during Golgi reformation.

Recently, NSF- and SNARE-mediated membrane fusion events were shown as essential steps during nucleus formation upstream of membrane flattening (Baur, Ramadan et al. 2007).

3.3 p97 is a negative regulator of Aurora B

We have found that p97 removes Aurora B from chromatin thereby inactivating it and promoting nucleus formation. In other words, p97 inactivates an inhibitor of nucleus formation. Therefore, we propose that p97 is a negative regulator of Aurora B. The evidences that support this hypothesis are the following. p97 activity is only required for nucleus formation *in vitro*, if Aurora B activity is present. Further, the activity of Aurora B on chromatin is elevated in the late mitotic *Xenopus laevis* egg extracts, if p97 is inactivated. During exit of mitosis in *Xenopus laevis* egg extracts, the dephosphorylation of the Aurora B substrate histone H3 is retarded in absence of p97 activity (Ramadan, Bruderer et al. 2007).
Additionally, analysis of the regulation of Aurora B by p97 in *Caenorhabditis elegans* confirmed the proposed model that p97 is a negative regulator of Aurora B. Knockdown of p97 in the background of a temperature sensitive Aurora B mutant worm results in a rescue of H3 phosphorylation on chromatin. This shows that reduced p97 activity stabilizes the temperature sensitive mutant of Aurora B. Additionally, knocking down p97 in temperature sensitive Aurora B mutant worms suppresses embryonic lethality. This indicates that p97 is an important negative regulator of Aurora B also *in vivo*.

3.4 p97 physically extracts Aurora B from chromatin

The chromosomal passenger complex binds to mitotic chromatin in prometaphase. Efficient chromatin binding of the chromosomal passenger complex requires Dasra proteins in *Xenopus laevis* egg extracts (Kelly, Sampath et al. 2007). Dasra and survivin can regulate Aurora B/INCENP localization to centromeres, whereas survivin may comprise the chromatin binding, but Dasra stabilizes the trimeric interaction Survivin/Dasra/INCENP, which confers interdependence (Vader, Kauw et al. 2006) and (Jeyaprakash, Klein et al. 2007). Neither the sequence or location where the chromosomal passenger complex associates to chromatin is known, nor is it known if substrates of Aurora B can recruit the chromosomal passenger complex to chromatin. Besides that, chromatin binding clusters Aurora B and activates it (Kelly, Sampath et al. 2007). Cytosolic Aurora B has low activity; therefore, important regulations of Aurora B may occur on chromatin.
We found that inactivation of p97 in *Xenopus laevis* egg extracts results in slightly elevated steady state levels of Aurora B on chromatin and elevated activity chromatin bound Aurora B

kinase activity, whereas no change of activity was detected in cytosol (Ramadan, Bruderer et al. 2007). This shows that p97 regulates an active, chromatin bound Aurora B pool. A model that explains the elevated Aurora B levels and activity on chromatin relies on the mobilization of Aurora B by p97 from chromatin. The mobilization model is supported by analogy to the p97 mechanism in the ERAD (Ye, Meyer et al. 2003) and in the activation of transcription factors (Rape, Hoppe et al. 2001), (Shcherbik and Haines 2007).

We show with a mobilization assay of Aurora B, that $p97^{Ufd1-Npl4}$ mobilizes chromatin bound, polyubiquitinated Aurora B in an energy dependent fashion. This finding strongly supports the proposed hypothesis that the function of p97 in the investigated process is analogous to its function in ERAD and transcription activation. Additionally, p97 binds Aurora B only in its polyubiquitinated forms. The same was reported also for other substrates of p97. $p97^{Ufd1-Npl4}$ binding to Mga2p120-Mga2p90 complexes requires polyubiquitination (Shcherbik and Haines 2007). Additionally, in ERAD a dual recognition mechanism has been proposed for substrate recognition, direct and via the polyubiquitin chains (Ye, Meyer et al. 2003). We could not detect binding of p97 to unmodified Aurora B. Hence, it would be a weak interaction, if existent. Newer data on the ERAD process suggest that the finding that p97 interacts with unmodified substrates might rather come from additional factors recruiting p97 to substrate ligase complexes at the membrane. As an example, ER membrane bound Ubx2 can recruit p97 to ligases and their substrates (Schuberth and Buchberger 2005). In the ERAD system, inactivation of p97 results in accumulation of polyubiquitinated substrates on the ER membrane (Ye, Meyer et al. 2001). Indeed, lysine-48 polyubiquitinated Aurora B can be detected on chromatin and the inhibition of p97 activity results in an accumulation of polyubiquitinated Aurora B on chromatin.

Aurora B mobilization as well as other substrate mobilization reactions by $p97^{Ufd1-Npl4}$ are ATP dependent processes. Therefore, we investigated the regulation of the ATPase activity of p97 by Ufd1-Npl4. It is known that p47 suppresses the ATPase activity of p97 by 85% upon binding to it (Meyer, Kondo et al. 1998). $p97^{Ufd1-Npl4}$ display equal ATPase activity as p97 alone, thus binding of Ufd1-Npl4 does not suppress p97 ATPase activity. For $p97^{p47}$ complexes no mobilization remodelling activity is reported. Thus, the inactivation of the ATPase activity of p97 in $p97^{p47}$ complexes could indicate a different mechanism. Interestingly, the difference in regulation of the ATPase activity is not mediated through the binding, since both protein have the same bipartite binding mechanism and a fragment of p47 containing only the binding module does not suppress p97 (Bruderer, Brasseur et al. 2004). Additional non binding domains in p47 must cause the suppression. Recently, it has been reported, that p47 is in a trimeric form in solution and that it binds in a trimeric state to the top of p97. Therefore, the trimerisation of p47 could contribute to the suppression of the ATPase activity of p97 by restricting the movements of p97 (Beuron, Dreveny et al. 2006). In the case of Ufd1-Npl4, no oligomerization is evident (Pye, Beuron et al. 2007). This could explain the observed differences between p47 and Ufd1-Npl4 in respect to the regulation of the ATPase activity of p97 (Figure 32).

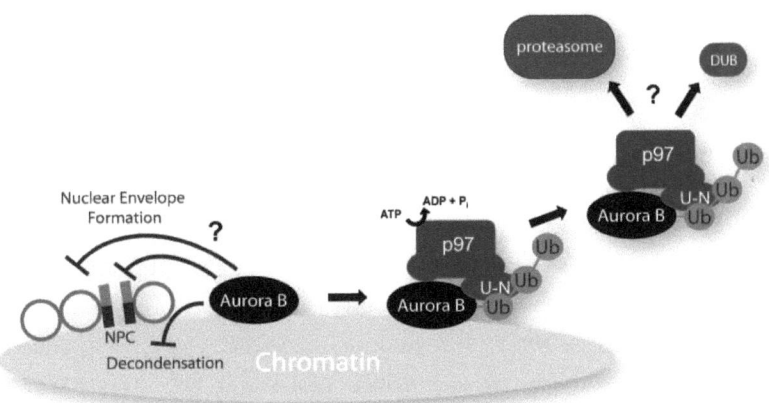

Fig. 32 p97 dependent mobilization of polyubiquitinated Aurora B from chromatin. $p97^{Ufd1-Npl4}$ complexes bind chromatin associated Aurora B after its polyubiquitination. Subsequently, the polyubiquitinated Aurora B is mobilized from chromatin. This inactivates the kinase activity of Aurora B and allows dephosphorylation of Aurora B substrates and promotes nuclear envelope formation around chromatin. The fat of the mobilized fraction of Aurora B remains unclear, indications point to proteasomal degradation.

The chromatin extraction model presented is at first view in contradiction to the following observation by Vong et al.. They reported that Ufd1 is involved in recruiting survivin, polyubiquitinated at lysine-63, to kinetochores in prometaphase and metaphase in somatic cells. Therefore, in absence of Ufd1 survivin levels drop on kinetochores (Vong, Cao et al. 2005). A difference to the finding of Vong et al. is that the embryonic system used here, is in interphase and has no centromeric Aurora B localization. Additionally, the polyubiquitination of Aurora B detected in the *Xenopus laevis* egg extract system was lysine-48 chains rather than lysine-63. Thus it may me a different process in a different system.

It remains unknown, what happens to the other chromosomal passenger proteins upon mobilization of Aurora B. Evidences suggest the existence of chromosomal passenger complexes of differing compositions. In immature G2 oocytes, only Aurora B, TD-60 and Dasra A are present, while little or no survivin or INCENP is present and Aurora B may not be in complex with TD-60 and Dasra A (Yamamoto, Lewellyn et al. 2008). It could be speculated that a chromosomal targeting complex of survivin and Dasra remains on chromatin while Aurora B is mobilized. Alternatively, Aurora B could be mobilized from INCENP, thus inactivating it (Sessa, Mapelli et al. 2005). A segregase activity has been published for p97$^{Ufd1-Npl4}$ during activation of the transcription factors Sp23/Mga2. Activation of the membrane bound transcription factors Sp23/Mga2 results in homo-dimerisation and processing of one moiety of the dimer to a fragment that has lost its trans-membrane domain. p97$^{Ufd1-Npl4}$ dissociates the processed, active transcription factor from its unprocessed partner. Subsequently, the mobilized active transcription factor can enter the nucleus and initiate transcription (Figure 7).

Recently the structure of survivin/Dasra/INCENP was solved. Survivin and Dasra bind to the N-terminus of INCENP. Dasra and INCENP associate with the C-terminal helical domain of survivin to form a tight three-helical bundle (Jeyaprakash, Klein et al. 2007) (Figure 11b). The related AAA-ATPase NSF$^{alpha-SNAP}$ dissociates cis-SNARE complexes by segregating highly stable four helix bundle complexes (Sollner, Bennett et al. 1993). Thus, it is tempting to speculate that p97$^{Ufd1-Npl4}$ may mobilize Aurora B with INCENP from chromatin, by dissociating the three helix bundle, thereby separating INCENP and Aurora B from survivin and Dasra.

On the other hand, there is also evidence that survivin may also be mobilized from chromatin. We have shown that in the absence of p97 activity the levels of polyubiquitinated forms of survivin and Aurora B increase in *Xenopus laevis* cytosol. This may indicate a common regulation and mobilization of Aurora B and survivin by p97. Additionally, it was observed in HeLa cells, that Aurora B as well as survivin persist on chromatin upon downregulation of Ufd1 and Npl4 (personal communication Dr. O. Popp). Additionally, Cullin3 dependent regulation of Aurora B on chromatin includes a similar regulation of survivin (Sumara, Quadroni et al. 2007). This model is supported by the finding that the removal of Aurora B from the clustered state on chromatin into the cytosol is enough to reduce Aurora B activity drastically (Kelly, Sampath et al. 2007).

In either case mobilization of Aurora B from chromatin inactivates it.

3. Discussion

The findings on the molecular mechanism of p97$^{Ufd1-Npl4}$ complexes presented here and published data from the ERAD as well as from transcription activation processes point to a general underlying mechanism in p97$^{Ufd1-Npl4}$ complex dependent processes. The energy dependent mobilization of an ubiquitinated factor from a structure is the general feature. It is reasonable to speculate that this may be the underlying mechanism of all p97$^{Ufd1-Npl4}$ processes. Additional other processing and recruiting factors may regulate and modify the general segregation process. Therefore, the fate of the mobilized factors may be variable.

It is not known how p97 exerts the remodelling activity on its substrates. It was suggested that p97 translocates its substrates through the pore in the middle of the hexamer. Central pore residues of p97 mediate the substrate binding of p97 required to ERAD substrates (DeLaBarre, Christianson et al. 2006). A threading mechanism has been reported for several AAA-ATPases like the bacterial chaperone ring complex associated coaxially with the proteolytic cylinder ClpA/ClpP (Reid, Fenton et al. 2001), for the proteasomal AAA-ATPases (reviewed in (Crews 2003)) and for the heat shock chaperone HSP104/ClpB (Weibezahn, Tessarz et al. 2004). However, the protein unfolding activity of these AAA+ ATPases is critically dependent on the presence of aromatic residues in the central pore. p97 contains aliphatic residues leucine and alanine of the D1 pore instead of aromatic tyrosine residues. Mutations of the aliphatic residues in aromatic can induce unfolding activity (Rothballer, Tzvetkov et al. 2007). Furthermore, the pore of p97 is narrow and reported to be clocked by a Zn^{2+} ion. Hence, it was speculated that complete unfolding of a substrate is desirable only for a proteolytic activity enabled by the AAA ATPase like in the case of the proteasome (DeLaBarre and Brunger 2003). This puts the substrate threading mechanism through the central pore of p97 into question.

Exploiting the homology of p97 and the proteasomal regulatory Rpt proteins, it was recently proposed, that p97 interacts with its substrates in the D2 pore. This model is supported by the finding that the D2 domain shows drastic conformational changes in the different nucleotide states. Additionally, D2 but not D1 ATP binding mutants trap ERAD substrates (DeLaBarre, Christianson et al. 2006).

It is also imaginable that p97 segregates substrates via the conformational changes of the N-domains (Beuron, Dreveny et al. 2006) including movements of bound substrate adaptors and thereby changing the conformation of the ubiquitinated substrate. However, the substrate adaptors have a relatively low affinity of ubiquitin, which makes unfolding of stable complexes unlikely.

It will be a great step forward in understanding of p97 dependent processes if the molecular mechanism of the substrate processing by p97 is solved.

3.5 Aurora B is a novel inhibitor/regulator of nucleus formation

We have shown for the first time that Aurora B regulates or inhibits nucleus formation late in mitosis. In *Xenopus laevis* egg interphase extracts, nuclei do not form if Aurora B activity is not properly regulated and has a too high activity. It is unclear how Aurora inhibits nucleus formation. From the literature, several Aurora B substrates and many other functions also earlier in mitosis are known; Aurora B is involved in the establishment of a stable bipolar spindle, chromatin remodelling, corrections of kinetochore-microtubule attachment errors via sensing the lack of tension and activation of the spindle assembly checkpoint, sister chromatid cohesion, regulation of mitotic progression, and cytokinesis (reviewed in (Ruchaud, Carmena et al. 2007)). Not much is known about the molecular details and requirements of these processes. It is reasonable to discuss if one of those processes may explain the role of Aurora B in nucleus assembly.

One important aspect of the role Aurora B is its involvement in chromatin remodelling during mitosis. Aurora B contributes to mitotic chromatin condensation (Vagnarelli and Earnshaw 2004), (Hagstrom, Holmes et al. 2002), (Kaitna, Pasierbek et al. 2002). Furthermore, Aurora B can mediate chromatin condensation in *Xenopus laevis* egg interphase cytosol in absence of phosphatase activity by addition of okadaic acid, this shows that still a residual Aurora B activity is present in *Xenopus laevis* egg interphase cytosol (Takemoto, Murayama et al. 2007). Maximal chromatin compaction is reached late in anaphase, after the segregation of sister chromatids. The anaphase compaction is dependent on Aurora B and dynamic microtubules and manifests in a shortening of the chromatid arms from telomere to centromere (Mora-Bermudez, Gerlich et al. 2007). Aurora B is responsible for targeting of a part of condensin I to chromatin in mitosis (Takemoto, Murayama et al. 2007) and (Lipp, Hirota et al. 2007). In addition to that, Aurora B phosphorylates histone H3 at serine 10 during mitosis, but the physiological role of H3 phosphorylation during mitosis is a longstanding question. Phosphorylation at serine 10 regulates indirectly the association of nuclear envelope membrane to chromatin. Tri-methylation of histone H3 lysine-9 is important to recruit heterochromatin protein 1 (HP1) to discrete regions on the chromatin. HP1 binds to LAP2beta and B-type lamins to the surface of chromatin, suggesting an involvement of HP1 proteins in nuclear envelope formation. If histone H3 is phosphorylated at serine 10 by Aurora B, HP1 does not bind to it (Hsu, Sun et al. 2000), (Fischle, Tseng et al. 2005), (Hirota, Lipp et al. 2005), (Kourmouli, Theodoropoulos et al. 2000).

These chromatin remodelling processes may antagonize nucleus formation. Condensed chromatin and a persisting spindle could interfere with nuclear envelope formation. The spindle regulation of Aurora B cannot explain its inhibition of nucleus formation since the inhibition of nucleus formation by p97/Aurora B in *Xenopus laevis* egg interphase cytosol is independent of the spindle (personal communication Kristijan Ramadan). The phosphorylation of H3 by Aurora B may indirectly inhibit nuclear envelope formation. This remains to be shown.

Another possibility is that Aurora B regulates nuclear envelope precursor membrane vesicles during exit of mitosis. Dephosphorylation of mitotic phosphorylation sites on membranes by PP1 in the synthetic phase cytosol fraction promote targeting of the membranes to chromatin (Ito, Koyama et al. 2007). Aurora B is in many cases an antagonist of PP1 (Emanuele, Lan et al. 2008), (Pinsky, Kotwaliwale et al. 2006), (Hsu, Sun et al. 2000), (Francisco and Chan 1994). Therefore, enhanced Aurora B activity could shift the equilibrium of phosphorylation/dephosphorylation to interfere with chromatin docking of membrane vesicles and stop nuclear envelope formation.

Alternatively, Aurora B may indirectly affect nuclear envelope formation by regulating NPCs, which are phosphorylated during mitosis (Margalit, Vlcek et al. 2005). Nuclear envelope formation and NPC assembly are connected during nucleus reformation. Depletion of the nucleoporin POM121 or NDC1 strongly inhibits nuclear membrane fusion and NPC assembly (Antonin, Franz et al. 2005) and (Mansfeld, Guttinger et al. 2006). Since the inhibitory effect of POM121 depletion can be overcome by co-depletion of Nup107, it was suggested that nuclear membrane fusion and NPC assembly are regulated by an assembly checkpoint. In analogy to the function of Aurora B in spindle checkpoint, Aurora B may cooperate in a NPC assembly and membrane fusion checkpoint to ensure proper sequential coordination of these processes late in mitosis. Thereby it would ensure faithful termination of mitosis.

The Ran pathway is thought to provide spatial control for NE formation. Generation of RanGTP on chromatin directs NPC assembly to chromatin. The Ran pathway also regulates NE membrane fusion, although the molecular basis is less well understood and may even be an indirect effect of NPC assembly regulation. Our new data now suggest that Aurora B and its regulation by p97 represent an element that introduces the spatial element also into regulation of NE formation by phosphorylation. Therefore, the p97/Aurora B and Ran pathways may functionally interact. Depletion of p97 in *C. elegans* embryos causes an increase of chromatin-associated Aurora B activity and ectopic assembly of nucleoporins in the cytoplasm, suggesting that overactivation of Aurora B on chromatin diverts NPC assembly away from chromatin and may thereby regulate nuclear envelope formation (Ramadan, Bruderer et al. 2007)

3.6 Fate of Aurora B after extraction from chromatin

In some systems, it has been found that Aurora B is ubiquitinated by the APC/C^{Cdh1} and degraded quantitatively by the proteasome at the end of mitosis (Stewart and Fang 2005), (Nguyen, Chinnappan et al. 2005). In contrast to that, Cdh1 is not present in *Xenopus laevis* egg extract (Lorca, Castro et al. 1998), and Aurora B not quantitatively degraded. Nevertheless, we detected of polyubiquitinated Aurora B, in *Xenopus laevis* egg cytosol as well as on chromatin. This interesting finding asks what the fate of the modified Aurora B is, whether it is degraded, deubiquitinated or regulated.

The inhibition of deubiquitinating enzymes, containing an active cysteine, gave rise to elevated levels of the total pool of polyubiquitinated proteins in *Xenopus laevis* egg cytosol, but did not result in raised levels of polyubiquitinated Aurora B, thus making the deubiquitination of Aurora B by deubiquitinating enzymes in this process unlikely. It remains the possibility that metallo-isopeptidases deubiquitinate Aurora B, which are not inhibited by the method used. Additionally, the levels of polyubiquitinated Aurora B rose in absence of the activity of the proteasome in *Xenopus laevis* egg cytosol. This shows that some polyubiquitinated Aurora B is subjected to proteasomal degradation. But it does not show, whether p97 is involved in the regulation of the fraction of polyubiquitinated Aurora B, that is degraded. Additionally, the reliability of proteasome inhibitors in *Xenopus laevis* extracts is not beyond any doubt, since very high concentrations are required, which might lead to artifacts.

In the p97 dependent mobilization of Aurora B, inhibition of the proteasome resulted in a retardation of mobilization of Aurora B from chromatin. This is in agreement to findings in the ERAD process. ERAD substrates are mainly degraded by the proteasome after extraction from the ER membrane. Proteasomal inhibition results in accumulation of degradation substrates bound to the ER membrane, polyubiquitinated as well as unmodified (Ward, Omura et al. 1995). This finding may point to the conclusion, that the mobilized Aurora B is degraded at least partly by the 26S proteasome. To additionally compare the two processes and elucidate the fate of Aurora B, it remains to investigate if also polyubiquitinated Aurora B accumulates on chromatin in absence of proteasomal degradation.

However, the fate of the mobilized factors seems also to vary. Other $p97^{Ufd1-Npl4}$ dependent processes are known, where lysine-48 modifications do not result in degradation. Deubiquitinating enzymes like OTU1 have important regulatory functions in those processes and may even rescue ubiquitinated substrates from degradation (Rumpf and Jentsch 2006) . During ERAD, polyubiquitinated substrates are degraded by the proteasome in the cytosol. Even in the ERAD process, regulations by deubiquitinating enzymes like Ataxin-3 (Zhong and Pittman 2006) and other regulating factors like SVIP are present (Ballar, Zhong et al. 2007). The $p97^{Ufd1-Npl4}$ activity in transcription activation results in liberation of an active transcription factor. It is notable to mention that lysine-48 chain ubiquitination is involved in this process (Shcherbik and Haines 2007).

3.7 A Cullin3-based ligase complex may cooperate with p97

Recently, it was shown that the ligase Cullin3 with its substrate adaptors KLHL9 and KLHL13 ubiquitinates Aurora B on mitotic chromatin in somatic cells during prometaphase. Inactivation of the Cullin3 based ligases results in a persistence of Aurora B and survivin on mitotic chromatin. Sumara et al. hypothesize that Cullin3 is involved in the removal of Aurora B and survivin from mitotic chromatin (Sumara, Quadroni et al. 2007). The total levels of Aurora B remain constant during prometaphase when the Cullin3 dependent ubiquitination occurs and drop quantitatively only at the end of mitosis when Cdh1 is active (Fang, Yu et al. 1999). Since Cullin3 generates lysine-48 chains on its substrates (Ribar, Prakash et al. 2007), and proteasomal inhibition mimics the defects caused by lack of Cullin3, it may be speculated a local degradation of a subpopulation of Aurora B early in mitosis may occur.

Sumara et al. detected a physical interaction of Cullin3 with Aurora B during mitosis. This physical interaction could also be detected in *Xenopus laevis* interphase cytosol. Additionally, p97 was also in complex with the ligase Cullin3. The binding and cooperation of other ligases with p97 has been shown to be of great importance for the p97 dependent process in ERAD (Gauss, Sommer et al. 2006). These findings led to the hypothesis that p97 and Cullin3 may cooperate in regulating Aurora B.

Thus, we speculated that Cullin3 might also ubiquitinate Aurora B later in mitosis during nucleus formation. Cullin3 dependent polyubiquitination of Aurora B in *Xenopus laevis* egg cytosol could only be detected in absence of p97 activity. This means that the accumulation of polyubiquitinated Aurora B above the basal level in absence of p97 is dependent on Cullin3. It is possible that the polyubiquitinated forms of Aurora B generated by Cullin3 are rapidly degraded or deubiquitinated in a p97 dependent manner. Therefore, an accumulation of Cullin3 dependent polyubiquitinated Aurora B did only occur in absence of p97 activity. The ligase responsible for the basal cytosolic ubiquitination remains obscure and a more thorough investigation of this process is required. Especially, it needs to be shown that the mobilization of polyubiquitinated Aurora B from chromatin is dependent on Cullin3.

Comparing the defects of loss of function of Cullin3 and the $p97^{Ufd1-Npl4}$ complexes reveals astonishing similarities early in mitosis of somatic cell. As mentioned above, RNAi of Cullin3 reveals that it is required for correct chromosome alignment in metaphase (Sumara, Quadroni et al. 2007). RNAi of Ufd1 or Npl4 results in similar defects of chromosome alignment in metaphase. A proper metaphase plate does not establish (personal communication O. Popp). The interesting phenocopy is confirmed of the regulation of Aurora B and survivin. In both cases a persistence of Aurora B and survivin on chromatin was observed in absence of Cullin3 or Ufd1-Npl4 function. Therefore, it is reasonable to speculate that the two pathways may cooperate in regulating the mitotic activities of Aurora B early in mitosis.

The influence of p97 on nucleus formation is more difficult to address in somatic cells, since the process cannot be analyzed in an isolated manner from the processes earlier in mitosis. The inhibition of the $p97^{Ufd1-Npl4}$ pathway shows already problems early in mitosis and the cells that go through mitosis have strange multi-lobed nuclei (personal communication O. Popp). Additionally, Cullin3 RNAi in HeLa cells results in multinucleated cells and no influence of Cullin3 on nucleus formation was reported. As mentioned above, in somatic cells Aurora B is degraded quantitatively at the end of mitoses. This is not the case in embryonic cells. It may be that the regulatory pathway postulated here is more important for nucleus formation in embryonic cells, since a high cytosolic pool of Aurora B is present even during the exit of mitosis. In somatic cells the pathway may be more important during prometa- and metaphase, than later during nucleus formation.

3.8 General relevance on p97, Cullin3 and Aurora B in mitosis

So far, most of the work on degradation of factors during mitosis has been done on quantitative degradation. Important quantitative degradations are contributed by the SCF and the APC/C. Transition of G_1 to S-phase is controlled by the activity of CDK inhibitors. Degradation of these inhibitors by SCF^{Skp2} promotes entry into S-phase (Bashir and Pagano 2004). The APC/C is active during mitosis and G_1. Activated APC/C^{cdc20} quantitatively degrades several important substrates including cyclin B and securin, which are essential for cell cycle progression. APC/C^{cdh1} is activated after the APC/C^{cdc20}, it has a broader substrate spectrum and takes over quantitative degradation of many mitotic factors including cyclin B at the end of mitosis and remains active throughout next G_1.

In contrast to these global degradations, the regulation of Aurora B by p97 is not global. A spatially restricted inactivation of Aurora B on chromatin occurs. The global levels do not change detectable at that time. The local regulation of Aurora B by p97 and Cullin3 precedes global Aurora B degradation by APC/C^{cdc20} at the end of mitosis. It is apparent that the activity of Aurora B needs to be regulated tightly during several stages of mitosis, since it is important for several essential functions.

At the end of mitosis, an ordered regulation/inactivation of mitotic factors is required, which cannot be satisfied by the global degradation via Cdh1. The activities of the mitotic kinases Plk1, Aurora A and Aurora B decrease at different time points. Therefore, additional local and temporal regulations/degradations are supposedly required besides Cdh1. The destruction of both cyclin A and cyclin B is carried out by the APC/C, the slight difference in time of degradation however is less understood and may indicate an undiscovered regulatory pathway (Fung and Poon 2005). Another example is the stabilization of cut1/separase and/or cut2/securin by p97 in addition to the degradation by APC/C^{cdc20} in the transition of metaphase to anaphase (Ikai and Yanagida 2006). It was reported that securin can by polyubiquitinated by Ufd2 in addition to the established polyubiquitination by the APC/C^{cdc20} (Spinette, Lengauer et al. 2004). Hence, it emerges that there exist redundant or additional regulation mechanisms at least for some mitotic key players, which are globally regulated by cyclin dependent kinases or the APC/C.

The phenotypes observed during metaphase upon inhibition of p97$^{Ufd1-Npl4}$ suggest that the p97/Aurora B pathway may have additional important functions earlier mitosis. Taking into account that p97 also regulates Cut1 raises the question if also other important factors are regulated by p97 during mitosis. Factors which already are assigned to a certain regulation may be subjected to additional regulation mechanisms. It even may be a general feature that additional mechanisms of regulation of key mitotic factors are existent besides the global ones to ensure the high fidelity of mitosis. Identification of these regulators, of which we propose p97 is one, could help to get a more comprehensive understanding of the precise regulation of local and temporal subpopulations of mitotic players.

4. Material and Methods

4.1 Preparation of crude and fractionated *Xenopus laevis* egg extract

To prepare extracts, female *Xenopus laevis* frogs were primed by injection of 50 units pregnant mare serum gonadotropin, (Calbiochem) dissolved in sterile water. 3-5 days after the first injection, frogs were injected with 1000-1500 units of human chorionic gonadotropine (Sigma). Frogs laid eggs 12-36 h after the second injection. During these period frogs were kept in tap water containing 100 mM NaCl.

Eggs were collected and once washed in 2% cysteine pH 8.0. They were dejelled by incubation in 200-300 ml 2% cysteine pH 8.0 for about 5 min. After 5 min the cysteine solution was poured off and eggs were washed three times in 0.25x Modified Ringer's solution (MMR) (1 M NaCl, 20 mM KCl, 10 mM $MgSO_4$, 25 mM $CaCl_2$, 5 mM HEPES-KOH pH 7.7) and three times in egg lysis buffer (ELB) (250 mM sucrose, 2.5 mM $MgCl_2$, 1mM DTT, 50 mM KCl, 10 mM HEPES-KOH pH 7.7). After the last washing step, eggs were transferred to polypropylene tubes using a wide-mouthed plastic pipette. All centrifugation steps were carried out at 4°C and all other steps on ice. Eggs were centrifuged for 5 min at 120g in a SW41 rotor. Extra buffer was removed from the top of the eggs and 5 mg/ml aprotinin (dilution 1:1000), 10 mM leupeptin, 20 mM cytochalasin B (dilution 1:1000), and 10 mg/ml cycloheximide (dilution 1:200) added on top of the eggs. Eggs were lysed in a SW41 rotor by centrifugation at 38.000g for 20 min. Crude extract was removed by piercing the tube with a syringe, aliquoted, snap frozen in liquid nitrogen and stored at −80°C.

For fractionation of the extract, crude extract was further centrifuged in a TLS-55 rotor at 200.000g, for 70 min. Cytosol was removed and again centrifuged in a TLS-55 rotor at 200.000g for 25 min. In the meantime, the light membrane fraction was diluted in about 1 ml ELB and underlaid with a sucrose cushion (0.42 g sucrose in ELB). Cytosol was aliquoted after the second centrifugation step, snap frozen, and stored at −80°C. Membranes were centrifuged at 21.500g for 20 min in a TLS-55 rotor. Supernatant was removed and the membrane pellet resuspended in 50 μl ELB. Membranes were also aliquoted, snap frozen and stored at −80°C.

4.2 Preparation of demembranated sperm chromatin

Testes of three male frogs were removed and overnight incubated in 1x MMR ((10x MMR: 1 M NaCl, 20 mM KCl, 10 mM $MgSO_4$, 25 mM $CaCl_2$, 5 mM HEPES-KOH pH7.7, 0.8 mM EDTA)) containing 10 U/ml human chorionic gonadotropine at 4°C. At the next day, testes were transferred to a 10 cm plastic dish and blood vessels and fat were removed with forceps without damaging the testis. Testes were transferred to a new 10 cm plastic dish and incubated for 5 min at RT in 10 ml cold 1x nuclei preparation buffer (NPB) plus protease inhibitors. Testes were transferred in a 1.5 ml Eppendorf tube and squashed with a pistil until no clumps were visible anymore. 1 ml cold 1x NPB (2x: 500 mM sucrose, 30 mM HEPES-KOH pH 7.7, 1 mM spermidinetrihydrochloride, 0.4 mM sperminetetrahydrochloride, 2 mM DTT, 2 mM EDTA pH 8.0) plus protease inhibitors was added and sperm suspension transferred to a piece of cloth of a first aid box placed into a funnel. Eppendorf tubes were washed with 1 ml cold 1x NPB and the buffer added to the sperm suspension. 1x NPB plus protease inhibitors was added until suspension dripped through the cloth in a 15 ml vial. The cloth was washed with additional buffer and the remaining solution pressed out by hand. At the end the vial contained about 14 ml sperm suspension. Sperm suspension was centrifuged at 3000 rpm for 10 min at 4°C. 1 ml of cold 1x NPB plus protease inhibitors was equilibrated to RT. After the centrifugation, the supernatant was removed; the sperm pellet was resuspended in 9 ml cold 1x NPB plus protease inhibitors and centrifuged as before. During the centrifugation step 10 mg/ml L-α-lysophosphatidylcholine (Sigma) dissolved in H_2O was prepared. After centrifugation, the supernatant was removed and the pellet resuspended in 1 ml 1x NPB plus protease inhibitors, equilibrated to RT. 50 μl of 10 mg/ml L-α-lysophosphatidylcholine were added and the suspension gently mixed. Sperm suspension was incubated for 5 min at RT. After 5 min, 10 ml cold 1x NPB plus protease inhibitors and 3% BSA was added to the suspension and the suspension centrifuged as before. After this centrifugation, the pellet should be looser than before and no longer have redness. The supernatant was removed, the pellet resuspended in 5 ml cold 1x NPB plus 0.3% BSA without protease inhibitors and again centrifuged at 3000 rpm for 10 min at 4°C. The supernatant was removed and the pellet resuspended in 500 μl 1x NPB plus 0.3% BSA and 30 % glycerol. Sperm suspension was divided in 50 μl aliquots and snap frozen in liquid nitrogen. Before use, sperms were diluted in sperm dilution buffer (110 mM potassium acetate, 50mM Tris-HCl pH7.5, 250 mM sucrose).

4.3 Nuclear envelope formation assay

A typical 20 µl nuclear formation reaction contained 12 µl crude extract, 2 µl 10x ATP-regenerating system (20 mM ATP, 100 mM creatine phosphate, 1 mM GTP, 0.2 mg/ml creatine kinase), 1 µl sperm (1000 spermheads/µl) and 4.5 µl buffer, recombinant proteins or antibodies. Reactions were incubated at 19°C for indicated times.
NEM-treatment was as described elsewhere (Macaulay and Forbes 1996).
For morphological analysis of nucleus formation, samples were formaldehyde-fixed in phosphate-buffered saline, stained with DAPI and $DiOC_6$, and visualized with a Leica SP2 AOPS confocal microscope. For quantification, at least 50 nuclei were scored visually using a Zeiss Axiovert 100TV epifluorescence microscope. Epifluorescence pictures of at least 50 chromatin particles per data point were taken at identical settings, and mean intensities quantified with NIH ImageJ software.

4.4 Ubiquitination analysis

Ubiquitination reactions were done in egg cytosol, supplemented with ATP-regenerating system and HA-ubiquitin at indicated concentrations. The reactions were incubated for 30 min at 19°C. The reactions were stopped by addition of fresh 5 mM N-ethylmaleimide.

4.5 Depletions and immunoprecipitations

Aurora B was immuno-depleted from cytosol using rabbit antibodies and the corresponding preimmune antibodies as control. Depletion of p97 was done as described before (Hetzer, Meyer et al. 2001). Native immunoprecipitations with the anti-Aurora B, anti-survivin, anti-INCENP, and anti-Ufd1 were performed in XB buffer and precipitates washed in WB (50 mM HEPES-KOH pH 7.4, 120 mM NaCl, 1 mM $MgCl_2$, 1 mM EGTA and 0.1% Triton X-100). For denaturing immunoprecipitations, samples were first adjusted to 1% SDS and then diluted to 0.1% SDS in phosphate-buffed saline containing 1% BSA and 1% Triton X-100 before subsequent immunoprecipitations.

4.6 Aurora B mobilization assay

For Aurora B mobilization assays, sperm chromatin was preincubated in cytosol at 19 °C for 20 min. The reaction was diluted in ELB and chromatin isolated through a 1.5 M sucrose cushion at 10'000 g for 10 min in a swing-out rotor and recovered. The chromatin was then re-incubated with manipulated cytosol as indicated. Subsequently, it was re-isolated by centrifugation at 10'000 g for 10 min in a swing-out rotor for Western blotting analysis. Detection was done using chemiluminescence and quantified directly with a AlphaInnotech Fluorchem 8900 imager.

4.7 Analysis of chromatin bound ubiquitinated proteins

For ubiquitination assays, chromatin was incubated in cytosol containing HA-ubiquitin for 30 min and isolated. Chromatin-associated proteins were eluted in high-salt buffer (600 mM NaCl, 10 mM HEPES, 0.5 % NP-40, pH 7.4) for 10 min on ice. The eluate was diluted tenfold in buffer (110 mM KCl, 10 mM HEPES, 0.5 % NP-40, 1 % BSA pH 7.4) and subjected to immunoprecipitations.

4.8 Malachite green ATPase Assay

The assays were performed using 2 _g of recombinant p97 preincubated without or with varying amounts of the p47, p47-(244–370) or Ufd1-Npl4 for 30 min on ice. Proteins were then incubated in 100 µl buffer (50 mM HEPES pH 7.4, 150 mM KCl, 2.5 mM MgCl2, 5% glycerol, and 2 mM ATP) for 40 min at 37 °C. 40 µl samples were taken at the start and the end of the reaction. Calibration was carried out by using an aqueous solution of KH_2PO_4. Quenching of color development was obtained by addition of 34% citrate. Formation of orthophosphate was measured using the malachite green color reagent according to the method by Lanzetta *et al.* (31).

4.9 Protein expression

All proteins were expressed in *E. coli*. Bacteria cultures were grown at 37°C until OD 0.6 was reached. Protein expression was induced by adding 0.2 mM IPTG. Proteins were expressed for 4 h at 37°C or overnight at 18°C. Bacteria cells were harvested by centrifugation at 4500 rpm for 15 min at 4°C. Afterwards, pellets were washed with 1x PBS and stored at –80°C or resuspended in lysis buffer and further processed for protein purification.

Purification of His-tagged p97

His-tagged p97 was expressed in *E. coli* for 4 hours at 37°C. Then the cells were harvested. The bacteria pellets were resuspended in lysis buffer (500 mM KCl, 100 mM Tris-HCl pH 7.4, 5 mM $MgCl_2$, 5% glycerol, 1 mM ATP, 2 mM 2-mercaptoethanol and 20 mM imidazole) plus protease inhibitors (0.1 mM PMSF and Complete Protease Inhibitor cocktail, Roche) and lysed using a cell homogenizer. All further steps were carried out on ice or at 4°C. The total cell lysate was centrifuged for 20 min at 17.500 rpm. After centrifugation, Ni^{2+}NTA-beads, washed in lysis buffer, were added to the supernatant (cleared lysate). Binding was carried out for 20 min. After binding, beads were sedimented by centrifugation at 1.000 rpm for 2 min and loaded in a column. Subsequently, beads were washed 4 times with washing buffer (150 mM KCl, 50 mM HEPES-KOH pH 7.4, 5 mM $MgCl_2$, 5% glycerol, 1 mM ATP, 2 mM 2-mercaptoethanol and 20 mM imidazole). After washing, proteins were eluted from the beads with washing buffer containing 350 mM imidazole in 250 µl fractions. Protein concentration of the single fractions was determined using BIO-RAD protein assay and fractions, containing the highest protein amount were pooled. The buffer of the pooled fraction was exchanged by gelfiltration.

Gel filtration chromatography was performed on an Äkta*FPLC* (Amersham) using a Sephacryl 300 gelfiltration column. The flow rate of all steps was 0.3 ml/min. The column was equilibrated with two column volumes storage buffer (20 mM HEPES-KOH pH 7.5, 150 mM KCl, 1 mM $MgCl_2$, 1 mM ATP, 2 mM 2-mercaptoethanol and 10% glycerol). Peak fractions of the Ni^{2+} eluate were pooled and centrifuged for 15 min at 13.200 rpm at 4°C. Maximal 0.75 ml of the Ni^{2+} eluate was applied to the column. Proteins were collected in 0.5 ml fractions.

Ion exchange chromatography was performed on an Äkta*FPLC* (Amersham) using a MonoQ column. The column was equilibrated in buffer A (150 mM KCl, 20 mM HEPES-KOH pH 7.4, 5 mM $MgCl_2$, 5% glycerol and 1 mM DTT), B (Buffer A with 1M KCl) and again A. Peak fractions of the gelfiltration chromatography were pooled and applied to the column. After binding, column was washed with 5 column volumes buffer A to remove unbound material. Proteins were eluted by performing a linear salt gradient (150 mM KCl to 1 M KCl) with buffer A and B. p97 eluted at a salt concentration of about 350 mM. p97 containing fractions were pooled and salt concentration was adjusted to 150 mM KCl by dilution in buffer A without salt. Subsequently, the proteins were concentrated using a Amicon Ultra filtration device with a 100.000 MWCO. The resulting fraction was aliquoted, snap frozen and stored at -80°C.

Purification of His-tagged p47 and His-tagged p47(244-370)

His-tagged p47 and His-tagged p47 (244-370) were expressed in *E. coli* for 4 hours at 37°C. Then the cells were harvested. The bacteria pellets were resuspended in lysis buffer (300 mM KCl, 50 mM Tris-HCl pH 7.4, 5% glycerol, 2 mM 2-mercaptoethanol and 20 mM imidazole) plus protease inhibitors (0.1 mM PMSF and Complete Protease Inhibitor cocktail, Roche) and lysed using a cell homogenizer. The following steps were carried out on ice or at 4°C. The total cell lysate was centrifuged for 30 min at 18.000 rpm. After centrifugation, Ni^{2+}NTA-beads, washed in lysis buffer, were added to the supernatant (cleared lysate). Binding was carried out for 30 min. After binding, beads were sedimented by centrifugation at 1.000 rpm for 2 min and loaded in a column. Subsequently, beads were washed 4 times with washing buffer (150 mM KCl, 25 mM HEPES-KOH pH 7.4, 5% glycerol, 2 mM 2-mercaptoethanol and 20 mM imidazole). After washing, proteins were eluted from the beads with washing buffer containing 350 mM imidazole in 250 µl fractions. Protein concentration of the single fractions was determined using BIO-RAD protein assay and fractions, containing the highest protein amount were pooled. The buffer of the pooled fraction was exchanged by gelfiltration. Gel filtration chromatography was performed using an Äkta*FPLC* (Amersham) with a Superdex 200 gelfiltration column. The flow rate of all steps was 0.3 ml/min. The column was equilibrated with two column volumes storage buffer (25 mM HEPES-KOH pH 7.5, 150 mM KCl, 1 mM $MgCl_2$, 1 mM ATP, 2 mM 2-mercaptoethanol and 10% glycerol). Peak fractions of the Ni^{2+} eluate were pooled and centrifuged for 15 min at 13.200 rpm at 4°C. Maximal

0.75 ml of the Ni^{2+} eluate was applied to the column. Proteins were collected in 0.5 ml fractions.

p97 purification from rat liver

p97 from rat liver was purified from rat liver cytosol. The first step was a 30% ammonium sulphate precipitation on ice over night. Precipitates were pelleted by centrifugation 30 min at 10.000g. Pellets were resuspended in buffer suitable for anion exchange chromatography (300 mM KCl, 20 mM Tris-HCl pH 7.4, 5 mM MgCl$_2$, 1 mM ATP and 1 mM DTT(dithiothreitol)). The resuspended proteins were applied to a preequilibrated Hitrap Q column. A linear gradient to 600 mM KCl was performed. p97 eluted at about 380 mM KCl. The peak fraction was analysed by SDS-Page and the fractions containing the most protein pooled. Next, gelfiltration with a Superdex200 column was performed in the buffer (20 mM HEPES-KOH pH 7.5, 150 mM KCl, 1 mM MgCl$_2$, 1 mM ATP, 2 mM 2-mercaptoethanol and 5% glycerol). Peak fractions of the Hitrap Q eluate were pooled and centrifuged for 15 min at 13.200 rpm at 4°C. Maximal 0.75 ml of the Hitrap Q was applied to the column. Proteins were collected in 0.5 ml fractions.

Purification of strep-tagged survivin

strep-tagged survivin was expressed in *E. coli* for at 18°C over night in LB media containing 0.1 µM ZnSO$_4$. Then the cells were harvested. The bacteria pellets were resuspended in lysis buffer (150 mM KCl, 50 mM HEPES-KOH pH 7.4, 5% glycerol and 1 mM dithiothreitol) plus protease inhibitors (0.1 mM PMSF and Complete Protease Inhibitor cocktail, Roche) and lysed using a cell homogenizer. The next steps were carried out on ice or at 4°C. The total cell lysate was centrifuged for 30 min at 17.500 rpm. After centrifugation, strep-Tactin Sepharose beads, washed in lysis buffer, were added to the supernatant (cleared lysate). Binding was carried out for 60 min. After binding, beads were sedimented by centrifugation at 1.000 rpm for 2 min and loaded in a column. Subsequently, beads were washed 4 times with lysis buffer. After washing, proteins were eluted from the beads with lysis buffer containing 10 mM biotin in 250 µl fractions. Protein concentration of the single fractions was determined using BIO-RAD protein assay and fractions, containing the highest protein amount were pooled. The buffer of the pooled fraction was exchanged by gelfiltration.

Gel filtration chromatography was performed using an Äkta*FPLC* (Amersham) with Superdex 200 gelfiltration column. The flow rate of all steps was 0.3 ml/min. The column was equilibrated with two column volumes storage buffer (20 mM HEPES-KOH pH 7.5, 150 mM KCl, 1 mM $MgCl_2$, 1 mM ATP, 2 mM 2-mercaptoethanol and 10% glycerol). Peak fractions of the Ni^{2+} eluate were pooled and centrifuged for 15 min at 13.200 rpm at 4°C. Maximal 0.75 ml of the Ni^{2+} eluate was applied to the column. Proteins were collected in 0.5 ml fractions. The peak fractions were pooled and used for subsequent anion exchange chromatography.

Ion exchange chromatography was performed on an Äkta*FPLC* (Amersham) using a MonoQ column. The column was equilibrated in buffer A (150 mM KCl, 20 mM HEPES-KOH pH 7.4, 5 mM $MgCl_2$, 5% glycerol and 1 mM DTT), B (buffer A with 1M KCl) and again B. The proteins were applied to the column. After binding, column was washed with 5 column volumes buffer A to remove unbound material. Proteins were eluted by performing a linear salt gradient (150 mM KCl to 1 M KCl) with buffer A and B. *strep*-tagged survivin eluted at a salt concentration of about 440 mM. *strep*-tagged survivin containing fractions were pooled and salt concentration was adjusted to 150 mM KCl by dilution in buffer A without KCl. Subsequently, the proteins were concentrated using an Amicon Ultra filtration device with a 10.000 MWCO. The resulting fraction was aliquoted, snap frozen and stored at -80°C.

Purification of HA-tagged ubiquitin

HA-tagged ubiquitin and mutated versions were expressed in *E. coli* for 4 hours at 37°C. Then the cells were harvested. The bacteria pellets were resuspended in lysis buffer (55 M H_2O) and lysed using a cell homogenizer. The total cell lysate was centrifuged for 30 min at 17.500 rpm. Next, an acid precipitation was performed using 3.5% perchloric acid. Bacterial protein aggregate to white colour. Subsequently, the precipitated proteins were spun down with 20min at 17.500 rpm. The supernatant was dialysed into 25 mM ammonium acetate HCl pH: 4.5.

Ion exchange chromatography was performed on an Äkta*FPLC* (Amersham) using a Hitrap SP column. The column was equilibrated in buffer A (25 mM ammonium acetate HCl pH: 4.5), B (250 mM ammonium acetate KOH pH: 7.6) and again B. The proteins were applied to the column. After binding, column was washed with 5 column volumes buffer A to remove unbound material. Proteins were eluted by performing a linear salt gradient (25 mM ammonium acetate to 250 M ammonium acetate) with buffer A and B. HA-tagged ubiquitin eluted at a salt concentration of about 90 mM. HA-tagged ubiquitin containing fractions were pooled and dialyzed in storage buffer (20 mM HEPES-KOH pH 7.5, 150 mM KCl and 1 mM dithiothreitol). Subsequently, the proteins were concentrated using an Amicon Ultra filtration device with a 5.000 MWCO. The resulting fraction was aliquoted, snap frozen and stored at -80°C.

His-tagged KLHL13 wt and KLHL13 FL173AA

His-tagged KLHL13 wt and KLHL13 FL173AA fragments were expressed in bacteria. The fragments Then the cells were harvested. The bacteria pellets were resuspended in lysis buffer (50 mM KCl, 20 mM HEPES-KOH pH 7.4, 5 mM $MgCl_2$, 5% glycerol, 1% triton X-100 and 2 mM 2-mercaptoethanol) and lysed using a cell homogenizer. The total cell lysate was centrifuged for 30 min at 17.500 rpm. Next the fragments were purified via Ni^{2+} affinity eluted with 250 mM imidazole, desalted into lysis buffer with econopac columns and analysed by SDS-PAGE. The purification resulted in reasonable pure protein

Ufd1-Npl4 protein complexes

Ufd1-Npl4 complexes were kindly provided by Hemmo Meyer.

4.10 SDS-PAGE and Western blotting

Proteins were separated by SDS-PAGE according the standard procedure running at constant currant flow.
Proteins were transferred from SDS-gels to nitrocellulose membranes high bond (Amersham) by wet blotting. Transfer was carried out at 300 mA for 1h on ice the transfer buffer was 1x SDS-running buffer with 20% methanol).

4.11 Antibodies

The following antibodies were used in this thesis. Antibodies were raised in rabbits against *Xenopus* GST-Aurora B, GST-survivin and an INCENP peptide (CSNRHHLAVGYGLKY). Antibodies to p97 (HME5) and Ufd1 (5E2) were described previously (Meyer, Shorter et al. 2000).

Properties of the antibodies used:

antibody	species	Dilution Western blot	source
anti-MCM3	rabbit	1:1000	kind gift of P. Jackson
anti-p97 HME 7	rabbit	1:1000	Meyer et al., 2000
anti-mouse IgG HRP	goat	1:10000	Bio-Rad, 6516
anit-rabbit IgG HRP	goat	1:10000	Bio-Rad, 6517
anti-His IgG	mouse	1:4000	Amersham, 27-4710-01
anti-RGS-His IgG	mouse	1:2000	Qiagen
Anti-Ufd1 5E2	mouse	n.d.	H. Meyer
Anti-survivin purified xSur1	rabbit	1:1000	R.M.Bruderer
Anti-Aurora B (serum 55)	rabbit	1:1000	R.M.Bruderer
Biotinylated Anti-Aurora B	rabbit	1:500	R.M.Bruderer
Anti-INCENP 1123	rabbit	1:1000	R.M.Bruderer
Anti-cullin3	rabbit	1:1000	kind gift of I.Sumara
Anti-KLHL13	rabbit	1:500	kind gift of I.Sumara
anti-HA (clone HA11)	mouse	1:1000	Covance, PEP-101P

Table 1 List of all antibodies used in this study

4.12 Affinity purification of anti-survivin antibodies

5 mg *strep*-tagged survivin purified by streptactin chromatography were coupled to Affi-Gel® 10 activated affinity media (BIO-RAD) pre-equilibrated with 1x PBS. Coupling was performed for 4 h at 4°C. Subsequently, the column was washed three times with cold 1x PBS and uncoupled survivin was removed by elution with 0.1 mM glycine HCl pH 2.8. Column was washed with 1x PBS.

For the affinity purification 7 ml anti-survivin serum, complement inactivated by incubation for 30 min at 56°C, was added to the column. Binding was carried out for 60 min. After washing, the column with 1x PBS, antibodies were eluted with 0.1 M glycine HCl pH 2.8. 200 µl fractions were immediately neutralized with 20 µl 1M Tris-HCl pH 7.4. Protein concentration was determined BIO-RAD protein assay and fractions containing the highest protein amount pooled. Buffer of the pooled fraction was exchange by dialysis into PBS.

4.13 Purification of anti-INCENP antibodies

3mg of the peptide used for immunization were dissolved in coupling buffer (50 mM Tris-KOH pH 8.5, 5 mM EDTA). Peptide was coupled for 45 min to sulpholink resin. Then it was washed to remove unbound peptide with coupling buffer. The unspecific binding sites of the resin were blocked by incubation with 50 mM L-cysteine HCL pH: 8.6 for 45 min. Next, it was washed with 1 M NaCl and 0.1 M glycine pH 2.8.

For affinity purification 5 ml anti-INCENP serum, complement inactivated by incubation for 30 min at 56°C, was added to the column. Binding was carried out for 60 min. After washing, the column with 1x PBS, antibodies were eluted with 0.1 M glycine pH 2.8. 200 µl fractions were immediately neutralized with 20 µl 1M Tris-HCl pH 7.4. Protein concentration was determined BIO-RAD protein assay and fractions containing the highest protein amount pooled. Buffer of the pooled fraction was exchange by dialysis into PBS.

4.14 Cloning

PCR reactions were performed using the PfuTurbo® DNA-polymerase (Stratagene). Restriction enzymes and T4-DNA-ligase were purchased from New England Biolabs and used according to the manufacturer's protocol. For point mutations or insertion of amino acid into plasmid DNA the QuikChange® site-directed mutagenesis kit (Stratagene) was used. DNA was extracted from agarose gels using the QIAquick Gel Extraction Kit (QIAGEN). Transfection of *E. coli* cells with plasmid-DNA was performed according to standard procedures. For plasmid preparations the QIAGEN Plasmid Maxi Kit and QIAGEN Spin Miniprep Kit (QIAGEN) were used.

survivin-strep S126E

Survivin-strep wt was cloned by side directed mutagenesis of the survivin-strep wt construct by usage of the QuickChange kit.

	construct	vector	template	primer (5`-3`)
1	pET23.X.survivin.strep.S126E	pET23 Novagen	pET23.X survivin.strep.	s CTATCGCAAATTTGAGACTGTTGTCCTCC as GGAGGACAACAGTCTCAAATTTGCGATAG

Table 2 Primers used to clone pET23.X.survivin.strep.S126E s: sense, as: antisense

HA-Ubiquitin K48R and K63R

pET.HA-Ubiquitin K48R and K63R were cloned by side directed mutagenesis of the pET.HA-ubiquitin wt construct by usage of the QuickChange kit.

	construct	vector	template	primer (5`-3`)
1	pET.HA-Ubiquitin K48R	pET23 Novagen	pET.HA-Ubiquitin wt	s CTGACTACAACATCCAGCGTGAGT-CAACCCTGCAC as GTGCAGGGTTGACTCACGCTGGATGTTGTAGTCAG
2	pET.HA-Ubiquitin K63R	pET23 Novagen	pET.HA-Ubiquitin wt	a CTGACTACAACATCCAGCGTGAGTCAACCCTG-CAC as GTGCAGGGTTGACTCACGCTGGATGTTG-TAGTCAG

Table 3 Primers used to clone pET.HA-Ubiquitin K48R and K63R. s: sense, as: antisense

Plasmids encoding His-p47, His-p47(244-370), His-p97, p97$_{\Delta D2}$, p97$_{\Delta D2\ K251A}$, HA-ubiquitin wt, GST-survivin and survivin-*strep,* and His-tagged KLHL13 wt and KLHL13 FL173AA have been available in the laboratory.

5. References

Allen, M. D., A. Buchberger, et al. (2006). "The PUB domain functions as a p97 binding module in human peptide N-glycanase." J Biol Chem 281(35): 25502-8.

Anderson, D. J. and M. W. Hetzer (2007). "Nuclear envelope formation by chromatin-mediated reorganization of the endoplasmic reticulum." Nat Cell Biol 9(10): 1160-6.

Antonin, W., C. Franz, et al. (2005). "The integral membrane nucleoporin pom121 functionally links nuclear pore complex assembly and nuclear envelope formation." Mol Cell 17(1): 83-92.

Auld, K. L., A. L. Hitchcock, et al. (2006). "The conserved ATPase Get3/Arr4 modulates the activity of membrane-associated proteins in Saccharomyces cerevisiae." Genetics 174(1): 215-27.

Ballar, P., Y. Zhong, et al. (2007). "Identification of SVIP as an endogenous inhibitor of endoplasmic reticulum-associated degradation." J Biol Chem 282(47): 33908-14.

Bashir, T. and M. Pagano (2004). "Don't skip the G1 phase: how APC/CCdh1 keeps SCFSKP2 in check." Cell Cycle 3(7): 850-2.

Baur, T., K. Ramadan, et al. (2007). "NSF- and SNARE-mediated membrane fusion is required for nuclear envelope formation and completion of nuclear pore complex assembly in Xenopus laevis egg extracts." J Cell Sci 120(Pt 16): 2895-903.

Beuron, F., I. Dreveny, et al. (2006). "Conformational changes in the AAA ATPase p97-p47 adaptor complex." EMBO J 25(9): 1967-76.

Bodoor, K., S. Shaikh, et al. (1999). "Sequential recruitment of NPC proteins to the nuclear periphery at the end of mitosis." J Cell Sci 112 (Pt 13): 2253-64.

Boeddrich, A., S. Gaumer, et al. (2006). "An arginine/lysine-rich motif is crucial for VCP/p97-mediated modulation of ataxin-3 fibrillogenesis." EMBO J 25(7): 1547-58.

Bolton, M. A., W. Lan, et al. (2002). "Aurora B kinase exists in a complex with survivin and INCENP and its kinase activity is stimulated by survivin binding and phosphorylation." Mol Biol Cell 13(9): 3064-77.

Bosu, D. R. and E. T. Kipreos (2008). "Cullin-RING ubiquitin ligases: global regulation and activation cycles." Cell Div 3: 7.

Briggs, L. C., G. S. Baldwin, et al. (2008). "Analysis of nucleotide binding to p97 reveals the properties of a tandem AAA hexameric ATPase." J Biol Chem.

Bruderer, R. M., C. Brasseur, et al. (2004). "The AAA ATPase p97/VCP interacts with its alternative co-factors, Ufd1-Npl4 and p47, through a common bipartite binding mechanism." J Biol Chem 279(48): 49609-16.

Buchberger, A., M. J. Howard, et al. (2001). "The UBX domain: a widespread ubiquitin-like module." J Mol Biol 307(1): 17-24.

Cao, K., R. Nakajima, et al. (2003). "The AAA-ATPase Cdc48/p97 regulates spindle disassembly at the end of mitosis." Cell 115(3): 355-67.

Chantalat, L., D. A. Skoufias, et al. (2000). "Crystal structure of human survivin reveals a bow tie-shaped dimer with two unusual alpha-helical extensions." Mol Cell 6(1): 183-9.

Chen, J., S. Jin, et al. (2003). "Survivin enhances Aurora-B kinase activity and localizes Aurora-B in human cells." J Biol Chem 278(1): 486-90.

Crews, C. M. (2003). "Feeding the machine: mechanisms of proteasome-catalyzed degradation of ubiquitinated proteins." Curr Opin Chem Biol 7(5): 534-9.

Davies, B. A., J. D. Topp, et al. (2003). "Vps9p CUE domain ubiquitin binding is required for efficient endocytic protein traffic." J Biol Chem 278(22): 19826-33.

Davison, D. B. and J. F. Burke (2001). "Brute force estimation of the number of human genes using EST clustering as a measure." IBM Journal or Research and Development 45(3/4): 439-449.

Dechat, T., A. Gajewski, et al. (2004). "LAP2alpha and BAF transiently localize to telomeres and specific regions on chromatin during nuclear assembly." J Cell Sci 117(Pt 25): 6117-28.

DeLaBarre, B. and A. T. Brunger (2003). "Complete structure of p97/valosin-containing protein reveals communication between nucleotide domains." Nat Struct Biol 10(10): 856-63.

DeLaBarre, B., J. C. Christianson, et al. (2006). "Central pore residues mediate the p97/VCP activity required for ERAD." Mol Cell 22(4): 451-62.

Deutscher, M. P., J. N. Abelson, et al. (1990). "Methods in Enzymology Series Volume 182: Guide to Protein Purification." Academic Press 182.

Dumitrescu, T. P. and W. S. Saunders (2002). "The FEAR Before MEN: networks of mitotic exit." Cell Cycle 1(5): 304-7.

Ellenberg, J., E. D. Siggia, et al. (1997). "Nuclear membrane dynamics and reassembly in living cells: targeting of an inner nuclear membrane protein in interphase and mitosis." J Cell Biol 138(6): 1193-206.

Emanuele, M. J., W. Lan, et al. (2008). "Aurora B kinase and protein phosphatase 1 have opposing roles in modulating kinetochore assembly." J Cell Biol **181**(2): 241-54.

Fang, G., H. Yu, et al. (1999). "Control of mitotic transitions by the anaphase-promoting complex." Philos Trans R Soc Lond B Biol Sci **354**(1389): 1583-90.

Fischle, W., B. S. Tseng, et al. (2005). "Regulation of HP1-chromatin binding by histone H3 methylation and phosphorylation." Nature **438**(7071): 1116-22.

Francisco, L. and C. S. Chan (1994). "Regulation of yeast chromosome segregation by Ipl1 protein kinase and type 1 protein phosphatase." Cell Mol Biol Res **40**(3): 207-13.

Fu, J., M. Bian, et al. (2007). "Roles of Aurora kinases in mitosis and tumorigenesis." Mol Cancer Res **5**(1): 1-10.

Funakoshi, M., T. Sasaki, et al. (2002). "Budding yeast Dsk2p is a polyubiquitin-binding protein that can interact with the proteasome." Proc Natl Acad Sci U S A **99**(2): 745-50.

Fung, T. K. and R. Y. Poon (2005). "A roller coaster ride with the mitotic cyclins." Semin Cell Dev Biol **16**(3): 335-42.

Galan, J. M. and R. Haguenauer-Tsapis (1997). "Ubiquitin lys63 is involved in ubiquitination of a yeast plasma membrane protein." EMBO J **16**(19): 5847-54.

Gant, T. M. and K. L. Wilson (1997). "Nuclear assembly." Annu Rev Cell Dev Biol **13**: 669-95.

Gauss, R., T. Sommer, et al. (2006). "The Hrd1p ligase complex forms a linchpin between ER-lumenal substrate selection and Cdc48p recruitment." EMBO J **25**(9): 1827-35.

Gerace, L. and B. Burke (1988). "Functional organization of the nuclear envelope." Annu Rev Cell Biol **4**: 335-74.

Glotzer, M. (2005). "The molecular requirements for cytokinesis." Science **307**(5716): 1735-9.

Haglund, K., P. P. Di Fiore, et al. (2003). "Distinct monoubiquitin signals in receptor endocytosis." Trends Biochem Sci **28**(11): 598-603.

Haglund, K. and I. Dikic (2005). "Ubiquitylation and cell signaling." EMBO J **24**(19): 3353-9.

Hagstrom, K. A., V. F. Holmes, et al. (2002). "C. elegans condensin promotes mitotic chromosome architecture, centromere organization, and sister chromatid segregation during mitosis and meiosis." Genes Dev **16**(6): 729-42.

Hampton, R. Y. (2002). "ER-associated degradation in protein quality control and cellular regulation." Curr Opin Cell Biol **14**(4): 476-82.

Han, J. K. and R. Nuccitelli (1990). "Inositol 1,4,5-trisphosphate-induced calcium release in the organelle layers of the stratified, intact egg of Xenopus laevis." J Cell Biol **110**(4): 1103-10.

Hauf, S., R. W. Cole, et al. (2003). "The small molecule Hesperadin reveals a role for Aurora B in correcting kinetochore-microtubule attachment and in maintaining the spindle assembly checkpoint." J Cell Biol **161**(2): 281-94.

Hershko, A. and A. Ciechanover (1998). "The ubiquitin system." Annu Rev Biochem **67**: 425-79.

Hershko, A. and I. A. Rose (1987). "Ubiquitin-aldehyde: a general inhibitor of ubiquitin-recycling processes." Proc Natl Acad Sci U S A **84**(7): 1829-33.

Hetzer, M., H. H. Meyer, et al. (2001). "Distinct AAA-ATPase p97 complexes function in discrete steps of nuclear assembly." Nat Cell Biol **3**(12): 1086-91.

Hetzer, M., H. H. Meyer, et al. (2001). "Distinct AAA-ATPase p97 complexes function in discrete steps of nuclear assembly." Nat Cell Biol **3**(12): 1086-91.

Heubes, S. and O. Stemmann (2007). "The AAA-ATPase p97-Ufd1-Npl4 is required for ERAD but not for spindle disassembly in Xenopus egg extracts." J Cell Sci **120**(Pt 8): 1325-9.

Hirota, T., J. J. Lipp, et al. (2005). "Histone H3 serine 10 phosphorylation by Aurora B causes HP1 dissociation from heterochromatin." Nature **438**(7071): 1176-80.

Honda, R., R. Korner, et al. (2003). "Exploring the functional interactions between Aurora B, INCENP, and survivin in mitosis." Mol Biol Cell **14**(8): 3325-41.

Hong, W. (2005). "SNAREs and traffic." Biochim Biophys Acta **1744**(3): 493-517.

Hoppe, T., K. Matuschewski, et al. (2000). "Activation of a membrane-bound transcription factor by regulated ubiquitin/proteasome-dependent processing." Cell **102**(5): 577-86.

Hsu, J. Y., Z. W. Sun, et al. (2000). "Mitotic phosphorylation of histone H3 is governed by Ipl1/aurora kinase and Glc7/PP1 phosphatase in budding yeast and nematodes." Cell **102**(3): 279-91.

Hurley, J. H., S. Lee, et al. (2006). "Ubiquitin-binding domains." Biochem J **399**(3): 361-72.

Huyton, T., V. E. Pye, et al. (2003). "The crystal structure of murine p97/VCP at 3.6A." J Struct Biol **144**(3): 337-48.

Ikai, N. and M. Yanagida (2006). "Cdc48 is required for the stability of Cut1/separase in mitotic anaphase." J Struct Biol **156**(1): 50-61.

Ito, H., Y. Koyama, et al. (2007). "Nuclear envelope precursor vesicle targeting to chromatin is stimulated by protein phosphatase 1 in Xenopus egg extracts." Exp Cell Res 313(9): 1897-910.

Jeyaprakash, A. A., U. R. Klein, et al. (2007). "Structure of a Survivin-Borealin-INCENP core complex reveals how chromosomal passengers travel together." Cell 131(2): 271-85.

Johnson, A. E. and M. A. van Waes (1999). "The translocon: a dynamic gateway at the ER membrane." Annu Rev Cell Dev Biol 15: 799-842.

Kaitna, S., P. Pasierbek, et al. (2002). "The aurora B kinase AIR-2 regulates kinetochores during mitosis and is required for separation of homologous Chromosomes during meiosis." Curr Biol 12(10): 798-812.

Kelly, A. E., S. C. Sampath, et al. (2007). "Chromosomal enrichment and activation of the aurora B pathway are coupled to spatially regulate spindle assembly." Dev Cell 12(1): 31-43.

Kincaid, M. M. and A. A. Cooper (2007). "ERADicate ER stress or die trying." Antioxid Redox Signal 9(12): 2373-87.

Kondo, H., C. Rabouille, et al. (1997). "p47 is a cofactor for p97-mediated membrane fusion." Nature 388(6637): 75-8.

Konturek, S. J., W. E. Schmidt, et al. (1987). "Valosin stimulates gastric and exocrine pancreatic secretion and inhibits fasting small intestinal myoelectric activity in the dog." Gastroenterology 92(5 Pt 1): 1181-6.

Kourmouli, N., P. A. Theodoropoulos, et al. (2000). "Dynamic associations of heterochromatin protein 1 with the nuclear envelope." EMBO J 19(23): 6558-68.

Lanzetta, P. A., L. J. Alvarez, et al. (1979). "An improved assay for nanomole amounts of inorganic phosphate." Anal Biochem 100(1): 95-7.

Li, J. J. and S. A. Li (2006). "Mitotic kinases: the key to duplication, segregation, and cytokinesis errors, chromosomal instability, and oncogenesis." Pharmacol Ther 111(3): 974-84.

Lipp, J. J., T. Hirota, et al. (2007). "Aurora B controls the association of condensin I but not condensin II with mitotic chromosomes." J Cell Sci 120(Pt 7): 1245-55.

Lohka, M. J. (1998). "Analysis of nuclear envelope assembly using extracts of Xenopus eggs." Methods Cell Biol 53: 367-95.

Lorca, T., A. Castro, et al. (1998). "Fizzy is required for activation of the APC/cyclosome in Xenopus egg extracts." EMBO J 17(13): 3565-75.

Macaulay, C. and D. J. Forbes (1996). "Assembly of the nuclear pore: biochemically distinct steps revealed with NEM, GTP gamma S, and BAPTA." J Cell Biol **132**(1-2): 5-20.

Maiorano, D., J. M. Lemaitre, et al. (2000). "Stepwise regulated chromatin assembly of MCM2-7 proteins." J Biol Chem **275**(12): 8426-31.

Mansfeld, J., S. Guttinger, et al. (2006). "The conserved transmembrane nucleoporin NDC1 is required for nuclear pore complex assembly in vertebrate cells." Mol Cell **22**(1): 93-103.

Margalit, A., S. Vlcek, et al. (2005). "Breaking and making of the nuclear envelope." J Cell Biochem **95**(3): 454-65.

Marshall, I. C. and K. L. Wilson (1997). "Nuclear envelope assembly after mitosis." Trends Cell Biol **7**(2): 69-74.

Masui, Y. and C. L. Markert (1971). "Cytoplasmic control of nuclear behavior during meiotic maturation of frog oocytes." J Exp Zool **177**(2): 129-45.

McCollum, D. (2005). "Cytokinesis: breaking the ties that bind." Curr Biol **15**(24): R998-1000.

Meyer, H. H., H. Kondo, et al. (1998). "The p47 co-factor regulates the ATPase activity of the membrane fusion protein, p97." FEBS Lett **437**(3): 255-7.

Meyer, H. H., J. G. Shorter, et al. (2000). "A complex of mammalian ufd1 and npl4 links the AAA-ATPase, p97, to ubiquitin and nuclear transport pathways." EMBO J **19**(10): 2181-92.

Meyer, H. H., Y. Wang, et al. (2002). "Direct binding of ubiquitin conjugates by the mammalian p97 adaptor complexes, p47 and Ufd1-Npl4." EMBO J **21**(21): 5645-52.

Moir, D., S. E. Stewart, et al. (1982). "Cold-sensitive cell-division-cycle mutants of yeast: isolation, properties, and pseudoreversion studies." Genetics **100**(4): 547-63.

Moldovan, G. L., B. Pfander, et al. (2007). "PCNA, the maestro of the replication fork." Cell **129**(4): 665-79.

Mora-Bermudez, F., D. Gerlich, et al. (2007). "Maximal chromosome compaction occurs by axial shortening in anaphase and depends on Aurora kinase." Nat Cell Biol **9**(7): 822-31.

Mosesson, Y., K. Shtiegman, et al. (2003). "Endocytosis of receptor tyrosine kinases is driven by monoubiquitylation, not polyubiquitylation." J Biol Chem **278**(24): 21323-6.

Mullally, J. E., T. Chernova, et al. (2006). "Doa1 is a Cdc48 adapter that possesses a novel ubiquitin binding domain." Mol Cell Biol **26**(3): 822-30.

Muller, J. M., J. Shorter, et al. (2002). "Sequential SNARE disassembly and GATE-16-GOS-28 complex assembly mediated by distinct NSF activities drives Golgi membrane fusion." J Cell Biol **157**(7): 1161-73.

Murnion, M. E., R. R. Adams, et al. (2001). "Chromatin-associated protein phosphatase 1 regulates aurora-B and histone H3 phosphorylation." J Biol Chem **276**(28): 26656-65.

Murray, A. W. (2004). "Recycling the cell cycle: cyclins revisited." Cell **116**(2): 221-34.

Nagiec, E. E., A. Bernstein, et al. (1995). "Each domain of the N-ethylmaleimide-sensitive fusion protein contributes to its transport activity." J Biol Chem **270**(49): 29182-8.

Newport, J. and W. Dunphy (1992). "Characterization of the membrane binding and fusion events during nuclear envelope assembly using purified components." J Cell Biol **116**(2): 295-306.

Nguyen, H. G., D. Chinnappan, et al. (2005). "Mechanism of Aurora-B degradation and its dependency on intact KEN and A-boxes: identification of an aneuploidy-promoting property." Mol Cell Biol **25**(12): 4977-92.

Nigg, E. A. (2001). "Mitotic kinases as regulators of cell division and its checkpoints." Nat Rev Mol Cell Biol **2**(1): 21-32.

Ogunjimi, A. A., D. J. Briant, et al. (2005). "Regulation of Smurf2 ubiquitin ligase activity by anchoring the E2 to the HECT domain." Mol Cell **19**(3): 297-308.

Ogura, T. and A. J. Wilkinson (2001). "AAA+ superfamily ATPases: common structure--diverse function." Genes Cells **6**(7): 575-97.

Ozkan, E., H. Yu, et al. (2005). "Mechanistic insight into the allosteric activation of a ubiquitin-conjugating enzyme by RING-type ubiquitin ligases." Proc Natl Acad Sci U S A **102**(52): 18890-5.

Peng, J., D. Schwartz, et al. (2003). "A proteomics approach to understanding protein ubiquitination." Nat Biotechnol **21**(8): 921-6.

Peters, J. M., M. J. Walsh, et al. (1990). "An abundant and ubiquitous homo-oligomeric ring-shaped ATPase particle related to the putative vesicle fusion proteins Sec18p and NSF." EMBO J **9**(6): 1757-67.

Philpott, A. and G. H. Leno (1992). "Nucleoplasmin remodels sperm chromatin in Xenopus egg extracts." Cell **69**(5): 759-67.

Pickart, C. M. (1997). "Targeting of substrates to the 26S proteasome." FASEB J **11**(13): 1055-66.

Pickart, C. M. (2001). "Mechanisms underlying ubiquitination." Annu Rev Biochem **70**: 503-33.

Pines, J. (2006). "Mitosis: a matter of getting rid of the right protein at the right time." Trends Cell Biol 16(1): 55-63.

Pinsky, B. A., C. V. Kotwaliwale, et al. (2006). "Glc7/protein phosphatase 1 regulatory subunits can oppose the Ipl1/aurora protein kinase by redistributing Glc7." Mol Cell Biol 26(7): 2648-60.

Pye, V. E., F. Beuron, et al. (2007). "Structural insights into the p97-Ufd1-Npl4 complex." Proc Natl Acad Sci U S A 104(2): 467-72.

Rabouille, C., H. Kondo, et al. (1998). "Syntaxin 5 is a common component of the NSF- and p97-mediated reassembly pathways of Golgi cisternae from mitotic Golgi fragments in vitro." Cell 92(5): 603-10.

Ramadan, K., R. Bruderer, et al. (2007). "Cdc48/p97 promotes reformation of the nucleus by extracting the kinase Aurora B from chromatin." Nature 450(7173): 1258-62.

Rape, M., T. Hoppe, et al. (2001). "Mobilization of processed, membrane-tethered SPT23 transcription factor by CDC48(UFD1/NPL4), a ubiquitin-selective chaperone." Cell 107(5): 667-77.

Rauh, N. R., A. Schmidt, et al. (2005). "Calcium triggers exit from meiosis II by targeting the APC/C inhibitor XErp1 for degradation." Nature 437(7061): 1048-52.

Reid, B. G., W. A. Fenton, et al. (2001). "ClpA mediates directional translocation of substrate proteins into the ClpP protease." Proc Natl Acad Sci U S A 98(7): 3768-72.

Ribar, B., L. Prakash, et al. (2007). "ELA1 and CUL3 are required along with ELC1 for RNA polymerase II polyubiquitylation and degradation in DNA-damaged yeast cells." Mol Cell Biol 27(8): 3211-6.

Rosasco-Nitcher, S. E., W. Lan, et al. (2008). "Centromeric Aurora-B activation requires TD-60, microtubules, and substrate priming phosphorylation." Science 319(5862): 469-72.

Rothballer, A., N. Tzvetkov, et al. (2007). "Mutations in p97/VCP induce unfolding activity." FEBS Lett 581(6): 1197-201.

Rothman, J. E. (1994). "Intracellular membrane fusion." Adv Second Messenger Phosphoprotein Res 29: 81-96.

Ruchaud, S., M. Carmena, et al. (2007). "Chromosomal passengers: conducting cell division." Nat Rev Mol Cell Biol 8(10): 798-812.

Rumpf, S. and S. Jentsch (2006). "Functional division of substrate processing cofactors of the ubiquitin-selective Cdc48 chaperone." Mol Cell 21(2): 261-9.

Sato, B. K. and R. Y. Hampton (2006). "Yeast Derlin Dfm1 interacts with Cdc48 and functions in ER homeostasis." Yeast 23(14-15): 1053-64.

Schmidt, W. E., V. Mutt, et al. (1985). "Valosin: isolation and characterization of a novel peptide from porcine intestine." FEBS Lett **191**(2): 264-8.

Schubert, U., L. C. Anton, et al. (2000). "Rapid degradation of a large fraction of newly synthesized proteins by proteasomes." Nature **404**(6779): 770-4.

Schuberth, C. and A. Buchberger (2005). "Membrane-bound Ubx2 recruits Cdc48 to ubiquitin ligases and their substrates to ensure efficient ER-associated protein degradation." Nat Cell Biol **7**(10): 999-1006.

Schuberth, C. and A. Buchberger (2008). "UBX domain proteins: major regulators of the AAA ATPase Cdc48/p97." Cell Mol Life Sci.

Schulman, B. A., A. C. Carrano, et al. (2000). "Insights into SCF ubiquitin ligases from the structure of the Skp1-Skp2 complex." Nature **408**(6810): 381-6.

Sessa, F., M. Mapelli, et al. (2005). "Mechanism of Aurora B activation by INCENP and inhibition by hesperadin." Mol Cell **18**(3): 379-91.

Shcherbik, N. and D. S. Haines (2007). "Cdc48p(Npl4p/Ufd1p) binds and segregates membrane-anchored/tethered complexes via a polyubiquitin signal present on the anchors." Mol Cell **25**(3): 385-97.

Shorter, J. and G. Warren (1999). "A role for the vesicle tethering protein, p115, in the post-mitotic stacking of reassembling Golgi cisternae in a cell-free system." J Cell Biol **146**(1): 57-70.

Shorter, J. and G. Warren (2002). "Golgi architecture and inheritance." Annu Rev Cell Dev Biol **18**: 379-420.

Sloper-Mould, K. E., J. C. Jemc, et al. (2001). "Distinct functional surface regions on ubiquitin." J Biol Chem **276**(32): 30483-9.

Snider, J. and W. A. Houry (2008). "AAA+ proteins: diversity in function, similarity in structure." Biochem Soc Trans **36**(Pt 1): 72-7.

Sollner, T., M. K. Bennett, et al. (1993). "A protein assembly-disassembly pathway in vitro that may correspond to sequential steps of synaptic vesicle docking, activation, and fusion." Cell **75**(3): 409-18.

Spinette, S., C. Lengauer, et al. (2004). "Ufd2, a novel autoantigen in scleroderma, regulates sister chromatid separation." Cell Cycle **3**(12): 1638-44.

Stewart, S. and G. Fang (2005). "Destruction box-dependent degradation of aurora B is mediated by the anaphase-promoting complex/cyclosome and Cdh1." Cancer Res **65**(19): 8730-5.

Sugiyama, K., K. Sugiura, et al. (2002). "Aurora-B associated protein phosphatases as negative regulators of kinase activation." Oncogene **21**(20): 3103-11.

Sumara, I., M. Quadroni, et al. (2007). "A Cul3-based E3 ligase removes Aurora B from mitotic chromosomes, regulating mitotic progression and completion of cytokinesis in human cells." Dev Cell **12**(6): 887-900.

Swaminathan, S., A. Y. Amerik, et al. (1999). "The Doa4 deubiquitinating enzyme is required for ubiquitin homeostasis in yeast." Mol Biol Cell **10**(8): 2583-94.

Takemoto, A., A. Murayama, et al. (2007). "Analysis of the role of Aurora B on the chromosomal targeting of condensin I." Nucleic Acids Res **35**(7): 2403-12.

Tan, A. L., P. C. Rida, et al. (2005). "Essential tension and constructive destruction: the spindle checkpoint and its regulatory links with mitotic exit." Biochem J **386**(Pt 1): 1-13.

Tan, J. M., E. S. Wong, et al. (2008). "Lysine 63-linked ubiquitination promotes the formation and autophagic clearance of protein inclusions associated with neurodegenerative diseases." Hum Mol Genet **17**(3): 431-9.

Trinkle-Mulcahy, L. and A. I. Lamond (2006). "Mitotic phosphatases: no longer silent partners." Curr Opin Cell Biol **18**(6): 623-31.

Trombetta, E. S. and A. J. Parodi (2003). "Quality control and protein folding in the secretory pathway." Annu Rev Cell Dev Biol **19**: 649-76.

Uchiyama, K., E. Jokitalo, et al. (2002). "VCIP135, a novel essential factor for p97/p47-mediated membrane fusion, is required for Golgi and ER assembly in vivo." J Cell Biol **159**(5): 855-66.

Uchiyama, K., E. Jokitalo, et al. (2003). "The localization and phosphorylation of p47 are important for Golgi disassembly-assembly during the cell cycle." J Cell Biol **161**(6): 1067-79.

Uchiyama, K. and H. Kondo (2005). "p97/p47-Mediated biogenesis of Golgi and ER." J Biochem **137**(2): 115-9.

Uchiyama, K., G. Totsukawa, et al. (2006). "p37 is a p97 adaptor required for Golgi and ER biogenesis in interphase and at the end of mitosis." Dev Cell **11**(6): 803-16.

Uhlmann, F. (2003). "Separase regulation during mitosis." Biochem Soc Symp(70): 243-51.

Urbe, S., L. J. Page, et al. (1998). "Homotypic fusion of immature secretory granules during maturation in a cell-free assay." J Cell Biol **143**(7): 1831-44.

Vader, G., J. J. Kauw, et al. (2006). "Survivin mediates targeting of the chromosomal passenger complex to the centromere and midbody." EMBO Rep **7**(1): 85-92.

Vader, G., R. H. Medema, et al. (2006). "The chromosomal passenger complex: guiding Aurora-B through mitosis." J Cell Biol **173**(6): 833-7.

Vagnarelli, P. and W. C. Earnshaw (2004). "Chromosomal passengers: the four-dimensional regulation of mitotic events." Chromosoma **113**(5): 211-22.

Vigers, G. P. and M. J. Lohka (1991). "A distinct vesicle population targets membranes and pore complexes to the nuclear envelope in Xenopus eggs." J Cell Biol **112**(4): 545-56.

Vong, Q. P., K. Cao, et al. (2005). "Chromosome alignment and segregation regulated by ubiquitination of survivin." Science **310**(5753): 1499-504.

Walther, T. C., A. Alves, et al. (2003). "The conserved Nup107-160 complex is critical for nuclear pore complex assembly." Cell **113**(2): 195-206.

Wang, Q., L. Li, et al. (2006). "Regulation of retrotranslocation by p97-associated deubiquitinating enzyme ataxin-3." J Cell Biol **174**(7): 963-71.

Wang, Q., C. Song, et al. (2003). "Hexamerization of p97-VCP is promoted by ATP binding to the D1 domain and required for ATPase and biological activities." Biochem Biophys Res Commun **300**(2): 253-60.

Wang, Y., A. Satoh, et al. (2004). "VCIP135 acts as a deubiquitinating enzyme during p97-p47-mediated reassembly of mitotic Golgi fragments." J Cell Biol **164**(7): 973-8.

Ward, C. L., S. Omura, et al. (1995). "Degradation of CFTR by the ubiquitin-proteasome pathway." Cell **83**(1): 121-7.

Weibezahn, J., P. Tessarz, et al. (2004). "Thermotolerance requires refolding of aggregated proteins by substrate translocation through the central pore of ClpB." Cell **119**(5): 653-65.

Wheatley, S. P., A. J. Henzing, et al. (2004). "Aurora-B phosphorylation in vitro identifies a residue of survivin that is essential for its localization and binding to inner centromere protein (INCENP) in vivo." J Biol Chem **279**(7): 5655-60.

Whiteheart, S. W., K. Rossnagel, et al. (1994). "N-ethylmaleimide-sensitive fusion protein: a trimeric ATPase whose hydrolysis of ATP is required for membrane fusion." J Cell Biol **126**(4): 945-54.

Wickner, W. and A. Haas (2000). "Yeast homotypic vacuole fusion: a window on organelle trafficking mechanisms." Annu Rev Biochem **69**: 247-75.

Xu, L., Y. Wei, et al. (2003). "BTB proteins are substrate-specific adaptors in an SCF-like modular ubiquitin ligase containing CUL-3." Nature **425**(6955): 316-21.

Yamamoto, T. M., A. L. Lewellyn, et al. (2008). "Regulation of the Aurora B Chromosome Passenger Protein Complex During Oocyte Maturation in Xenopus." Mol Cell Biol.

Yang, L., T. Guan, et al. (1997). "Integral membrane proteins of the nuclear envelope are dispersed throughout the endoplasmic reticulum during mitosis." J Cell Biol **137**(6): 1199-210.

Ye, Y. (2006). "Diverse functions with a common regulator: ubiquitin takes command of an AAA ATPase." J Struct Biol **156**(1): 29-40.

Ye, Y., H. H. Meyer, et al. (2001). "The AAA ATPase Cdc48/p97 and its partners transport proteins from the ER into the cytosol." Nature **414**(6864): 652-6.

Ye, Y., H. H. Meyer, et al. (2003). "Function of the p97-Ufd1-Npl4 complex in retrotranslocation from the ER to the cytosol: dual recognition of nonubiquitinated polypeptide segments and polyubiquitin chains." J Cell Biol **162**(1): 71-84.

Ye, Y., Y. Shibata, et al. (2004). "A membrane protein complex mediates retro-translocation from the ER lumen into the cytosol." Nature **429**(6994): 841-7.

Zhong, X. and R. N. Pittman (2006). "Ataxin-3 binds VCP/p97 and regulates retrotranslocation of ERAD substrates." Hum Mol Genet **15**(16): 2409-20.

Zhong, X., Y. Shen, et al. (2004). "AAA ATPase p97/valosin-containing protein interacts with gp78, a ubiquitin ligase for endoplasmic reticulum-associated degradation." J Biol Chem **279**(44): 45676-84.

Abbreviations

AAA	ATPase associated with various cellular activities
ACP/C	anaphase promoting complex / cyclosome
ATP	adenosine 5`-triphosphat
AMP	adenosine 5`-monophosphat
BIR	baculovirus IAP repeats
BS1	binding site 1
BSA	bovine serum albumin
CDC	cell division cycle
Cdk	cyclin dependent kinase
CSF	cytostatic factor (CSF extract: cytostatic factor arrested extract)
$DiOC_6$	3,3`-dihexyloxacarbocyanine iodide
DTT	dithiothreitol
DNA	desoxyribonuclei cacid
DMSO	dimethylsulfoxid
ELB	egg lysis buffer
EGTA	ethylene glycol tetraacetic acid
ER	endoplasmic reticulum
ERAD	endoplasmic reticulum associated degradation
FEAR	CDC fourteen early anaphase release
FPLC	fast protein liquid chromatograpy
GDP	guanosine 5`-diphosphat
GPI	glycosylphosphatidylinisotol
GST	glutathione S-transferase
GTP	guanosine 5`-triphosphat
HA-tag	hemagglutinin epitope tag (YPYDVPDYA)
HECT	homologous to the E6-AP Carboxyl Terminus
HEPES	4-(2-hydroxyethyl)-1-piperazineethanesulfonic acid
kDa	kilo Dalton
INCENP	innercentromeric protein

6. Abbreviations

INM	inner nuclear membrane
KLHL	Kelch like
LAP2	lamina-associated polypeptide 2
LBR	lamin B receptor
ONM	outer nuclear membrane
OTU	ovarian tumour
MEN	mitotic exit network
MCAK	mitotic centromere-associated kinesin
mDa	mega Dalton
MG132	Z-Leu-Leu-Leu-aldehyde
MMR	Modified Ringer's solution
MWCO	molecular weight cut off
NE	nuclear envelope
NEM	N-ethylmaleimide
NIMA	never in mitosis gene a
NPB	nuclei preparation buffer
NPC	nuclear pore complex
NPL	nuclear protein localization
NSF	NEM-sensitive factor
Nup	nucleoporin
OLE	oleic acid
PAGE	polyacryl gel electrophoresis
PBS	phosphate buffered saline
PCNA	proliferating cell nuclear antigen
PCR	polymerase chain reaction
PFA	paraformaldehyde
PUB	PNGase (peptide N-glycosidase)/ubiquitin-associated
RNA	ribonucleic acid
RT	room temperature
SNAP	soluble NSF attachment protein
SNARE	soluble NSF attachment protein receptor
SPT	suppressor of Ty genes

6. Abbreviations

SRH	second region of homology
strep-tag	tag with affinity for streptactin (AWRHPQFGG)
SDS	sodium dodecylsulfate
TD60	telophase disc 60 protein
Tris	tris(hydroxymethyl)-aminomethane
UBA	ubiquitin association domain
UBD	ubiquitin fold domain
UBX	ubiquitin regulatory X
UFD	ubiquitin fusion degradation
UIM	ubiquitin interacting motif
UPS	ubiquitin proeatome system
UPR	unfolded protein responce
VBM	VCP-binding motif
VIM	VCP-interacting motif
VCP	valosin containing protein
wt	wild type

Acknowledgement

I want to thank Hemmo for the opportunity to do my PhD in his laboratory. It was a nice, intensive and instructive time.

I thank Matthias Peter and Jörg Höhfeld for being my committee members and Rudi Glockshuber for chairing my PhD defence.

I thank Peter Jackson for providing me antibodies.

I thank Kristijan for a delightful collaboration and for sharing his wisdom of life with me.
I thank Oli for a great collaboration and excursions in the Bavarian culture.

I thank Tina for excellent collaboration and for all their help in using the *Xenopus laevis* system.

Thanks go also to Toni and Sonja, who took care of the frogs and made life a lot easier for us.

I thank Tina and Christoph for correcting my thesis.

Furthermore, thanks to all members of the lab for the great time and the many BBQs and JumboJumbos as well as the great scientific support.

Special thanks goes to Catherine for the support in the laboratory and all proteins she purified for me.

Finally, I great thanks go to my family, who supports me all the time in a great way.

Last but not least, heartily thanks go to my Nicole, who had to sustain a sometimes scatterbrained me, still supports me and greatly distracts me from work with many adventures.

Thank you.

Die VDM Verlagsservicegesellschaft sucht für wissenschaftliche Verlage abgeschlossene und herausragende

Dissertationen, Habilitationen, Diplomarbeiten, Master Theses, Magisterarbeiten usw.

für die kostenlose Publikation als Fachbuch.

Sie verfügen über eine Arbeit, die hohen inhaltlichen und formalen Ansprüchen genügt, und haben Interesse an einer honorarvergüteten Publikation?

Dann senden Sie bitte erste Informationen über sich und Ihre Arbeit per Email an *info@vdm-vsg.de*.

Sie erhalten kurzfristig unser Feedback!

VDM Verlagsservicegesellschaft mbH
Dudweiler Landstr. 99
D - 66123 Saarbrücken
www.vdm-vsg.de

Telefon +49 681 3720 174
Fax +49 681 3720 1749

Die VDM Verlagsservicegesellschaft mbH vertritt

Printed by Books on Demand GmbH, Norderstedt / Germany